APPLICATION OF GIS AND REMOTE SENSING IN ENVIRONMENTAL MANAGEMENT

APPLICATION OF GIS AND REMOTE SENSING IN ENVIRONMENTAL MANAGEMENT

By

S.A. Abbasi

K.B. Chari

Centre for Pollution Control and Energy Technology
Pondicherry University
Pondicherry–605 014

DISCOVERY PUBLISHING HOUSE
NEW DELHI-110002

Edition - 2020

ISBN: 978-81-7141-963-0

Application of GIS and Remote Sensing in Environmental Management

Published by:
DISCOVERY PUBLISHING HOUSE PVT. LTD.
4383/4B, Ansari Road, Darya Ganj
New Delhi-110 002 (India)
Phone: +91-11-23279245, 23253475; 43596065
E-mail: discoverybooksindia@gmail.com
discoverypublishinghouse@gmail.com
web: www.discoverypublishinggroup.com

Printed at:
Infinity Imaging Systems
Delhi

Dedicated

To

The Loving and Caring Nandeshwar *ji*
(Dr. M.D. Nandeshwar, Scientist-F, CWRDM, Kozhikode), Padma *Bhabhi*, Anshu, and Mona

—Abbasi

My

Parents

Sri K. Yadagiri and Smt. K. Vijayalakshmi
Ever Comforting, Ever Inspiring

—Chari

डॉ. अरूण निगवेकर
अध्यक्ष

Dr. Arun Nigavekar
Chairman

विश्वविद्यालय अनुदान आयोग
बहादुर शाह ज़फर मार्ग, नई दिल्ली-110 002
UNIVERSITY GRANTS COMMISSION
BAHADUR SHAH ZAFAR MARG
NEW DELHI - 110 002
OFF : (011) 23239628
(011) 23221313
FAX : (011) 23231797
E-mail : narun42@hotmail.com

Foreword

Wetlands are among the most interesting and challenging entities from research point of view as they are very productive and biologically rich eco-systems on the earth. Today because of industrialization and enormous urbanization wetlands are amongst the most endangered entities. There is a need for eco-restoration and management of wetlands. It is very interesting subject and it throws challenge to the researchers. In recent time Geographic Information System (GIS) has proved to be a very useful technique and tool to study wetland management. Today in a true sense use of GIS for evaluating eco-systems of the wetlands has become an interdisciplinary subject, which requires expertise in various subjects and disciplines. It is very encouraging to see that the authors Prof. S.A. Abbasi and Prof. K.B. Chari has presented a very useful and detailed information on application of GIS and remote sensing in critical understanding of wetlands. They have presented intensive study on Kaliveli Wetlands. The book is very rich in data and has several illustrations in support of the information presented in the text. I think this would be a very rich material for all those who are interested in using GIS and remote sensing in understanding of wetlands. I would like to congratulate the authors for bringing out such a useful book on a topic immediate interest in today's world.

(Arun Nigavekar)
Chairman

Preface

Wetlands can veritably be called 'the cradles of life' because life began in water and is sustained by it. There are organisms which can survive without air to breathe but no organism, big or small, can survive without water.

Wetlands–rivers, lakes, ponds–have also been the nuclei around which civilizations have taken root and grown throughout human history. When wetlands faced a crisis, so did the people. It is no wonder that all civilizations treat water with respect, most even worship it!

And yet, strangely and very sadly–we have treated wetlands with indifference rather than care, cruelty rather than affection, contempt rather than respect. More 'developed', and 'advanced' an area, more pathetic is the condition of its wetlands. In cities, which are the epitomes of 'development', wetlands have either been dead and buried or tottering on the brink. Wherever urbanization reaches the shores of a wetland the countdown for the wetland's demise begins. Indeed land sharks do everything possible to hasten the death of a wetland so that they could grab the land that becomes available, if possible for putting up skyscrapers.

Such mishandling of nature's bounties can't go on without backlashes. It might take time for the backlashes to become strong enough to sting but when the sting begins it keeps on increasing with frightening severity. And all those cities and towns, all those people, who had harmed their wetlands are beginning to pay the price in terms of shortage of water, flooding (when it rains), pollution, and a ravaged microclimate. With the passage of time, and with demographic pressure forever getting heavier, the crisis is deepening by the hour. It, therefore, becomes imperative that restoration and conservation of wetlands is accorded very high priority.

Kaliveli, situated about 85 Km south of Chennai, close to Pondicherry, is one of the largest wetlands of Peninsular India. It has a water spread in excess of 70 Km^2 and a catchment over 10 times larger. It plays a crucial role in recharging groundwater, controlling microclimate, providing fisheries, serving as niche to numerous flora and fauna, cushioning the impact of cyclones, and providing numerous other benefits. Kaliveli is also a major wintering ground for migratory birds, hosting between 70,000 to 1,40,000 ducks, shorebirds, and terns every year !

Acknowledging the importance of Kaliveli, the state of neglect it has been going through, and the expertise available with the Centre for Pollution Control & Energy Technology, Pondicherry University, the University Grants Commission had sponsored a major R&D project on the ecology and conservation of Kaliveli. By applying the techniques of GIS (geographical information systems) and remote sensing to very extensive ground-truth studies, the project investigators have striven to create a body of knowledge which is applicable to Kaliveli but which, hopefully, goes beyond its immediate sphere of applicability and can serve as a do-how manual for future studies on other wetlands.

The authors express their deep sense of gratitude to the UGC for making this study possible and to Prof Arun Nigavekar for kindly agreeing to write what has turned out to be a highly perceptive *foreword*.

–SAA

–KBC

Contents

SECTION—I

INTRODUCTION

1

Introduction

Among the techniques and tools capable of assisting in resources identification, mapping, and utilization, GIS (Geographic Information System) has had the most spectacular growth. Within a few years of its formal introduction, GIS has come to influence each and every dimension of resource science.

GIS had acquired tremendous power and reach when it was used in tandem with remote sensing. Now, as newer advancements occur in the fields of information technology and communication engineering, the value and impact reach of GIS proportionately increase.

GEOGRAPHIC INFORMATION SYSTEMS

Geographic Information Systems (GIS) are designed to accept, organize, statistically analyze, and display diverse types of spatial data. These aspects are digitally referenced to a common coordinate system of particular projection and scale. Burrough and Mc Donnell (1998) has categorized various definitions of GIS based on their utility and function:

(a) Tool Box-based Definitions

(*i*) 'A powerful set of tools for collecting, storing, retrieving, at will, transforming anu displaying spatial data from the real world' (Burrough, 1986).

(ii) 'A system for capturing, storing, checking, manipulating, analyzing and displaying data which are specially referenced to the Earth' (Department of Environment, 1987).

(iii) 'An information technology which stores, analyses, and displays both spatial and non-spatial data' (Parker, 1988).

(b) Database Definitions

(i) 'A database system in which most of the data are spatially indexed and upon which a set of procedures operated in order to answer queries about spatial entities in the database' (Smith et. al., 1987).

(ii) 'Any manual or computer based set of procedures used to store and manipulate geographically referenced data' (Aronoff, 1989)

(c) Organization Based Definitions

(i) 'An automated set of functions that provides professionals with advanced capabilities for the storage, retrieval, manipulation and display of geographically located data' (Ozemoy et. al., 1981).

(ii) 'An institutional entity, reflecting on organizational structure that integrates technology with a database expertise and continuing financial support over time' (Carter, 1989).

(iii) 'A decision support system involving the integration of spatially reference data in a problem solving environment' (Cowen, 1988).

Geographic Information System (GIS) can also be defined as a set of integrated activities which provides us a tool to:

- Integrate geographic data received from different sources such as maps, charts, tables, aerial photographs, satellite imagery, GPS (Global Positioning System) in digital environment
- Attach thematic/attribute information to the geographic details

- Analyze results and build up queries based on spatial and/or attribute information
- Get the results in a desired form.

The word geographic implies that locations of the data items are known, or can be calculated, in terms of geographic coordinates (latitude, longitude). A Geographic Information System, then according to these criteria may be summarized as having the following characteristics (Martin, 1996):

Geographic: The system is concerned with data relating to geographic scales of measurement, and which are referred by some coordinate system to locations on the surface of the earth. Other types of information system may contain details about location, but here spatial objects and other locations are the very building blocks of the system.

Information: It is possible to use the system to ask questions of the geographic database obtaining information about the geographic world. This represents the extraction of specific and meaningful information from a diverse collection of data, and is only possible way in which the data are organized into a 'model' of the real world.

System: This is the environment, which allows data to be managed and questions to be posed. In the most general sense, a GIS need not be automated (a non-automated example would be a traditional map library), but should be an integrated set of procedures for the input, storage, manipulation and output of geographic information. Such a system is most readily achieved by automated means, and our concern here will be specifically with automated systems.

As suggested by the last point, the data in a GIS are subject to a series of transformations and may often be extracted or manipulated in a very different form to that in which they were collected and entered. This idea of a GIS as a tool for transforming spatial data is consistent with the traditional view of cartography. Thus, GIS have functional capabilities for data capture, input, manipulation, transformation, visualization, combination, query, analysis, modelling and output.

Analog Maps Vs Digital Maps

The conventional maps drawn or printed on paper strive to display information with reference to location. For example, a political map tells us which country, state, or city is located where and a soil map tells us the areas where different types of soils occur. The digital maps in GIS also handle data with reference to space but whereas the conventional maps are static and two dimensional, GIS is dynamic and multifaceted. GIS can not only display information with reference to space but also with reference to time. And GIS can also store, analyze, check, manipulate, and represent data besides displaying it.

Contemporary GIS

Earlier, the GIS development had been focused particularly on the use of networked workstations running under variants of the UNIX operating system. Numerous hardware manufacturers were involved in this field, and the leading software would generally be implemented under the major hardware suppliers systems. Common examples of such machine series at this level were Sun SPARC stations, Silicon Graphics, IBM RS 6000, and Digital DEC stations. Earlier, GIS, because of their heavy use of interactive graphics and database access, would not sit comfortable on mainframe machines running many other tasks simultaneously. Also, GIS have database requirements which until very recently have been too large for most PCs, hence the networked workstation environment was well suited for yesteryear's GIS applications.

In recent times most of the GIS companies are focusing on the 'desk top GIS' market. With the rapid advancements in the information technology and as well the ever reducing costs of computer hardware, the PC's have become very popular. As the processing power has become more widely available and GIS more widely known, the successively smaller organizations are making use of GIS.

The GIS companies are now trying to grab desktop GIS market by trimming, and redesigning their software into several upgradeable modules. Arguably the most popular of the GIS market leaders—ESRI's—has redesigned and repacked Arc Info, as Arc GIS. The Arc GIS now contains several hierarchical modules: (i) Arc View—the entry point module into Arc GIS which provides core mapping and GIS functionality (ii) Arc Editor—includes the functionalities of Arc View and bundles some more functionalities such as multi-user editing, versioning, custom feature classes, and dimensioning (iii) Arc Info—the top of in the line of Arc GIS package—can perform GIS data creation, querying, mapping, and analysis.

GIS has the potential to change the geological workplace drastically. As the personal computer has virtually eliminated the typewriter, automated GIS has almost replaced the light table and the map cabinet.

COMPONENTS OF GIS

There are four integrated components of GIS: data and databases, hardware, software including database management systems, and users (Fig. 1.1).

1. Data and Database

The data in a GIS are by definition geographic. Spatial data is the information pertaining to where the objects of interest are located. This information can be about the distribution and extent, adjacency, proximity and connectivity, attribute data, or observations about features.

2. Hardware

A fully functional GIS must contain hardware to support data input, output, storage, retrieval, display, and analysis.

Input Devices

(a) Analog data such as maps, satellite images and photographs need to be converted to digital data, in raster format or vector format, for which a scanner or/ and digitizer is required.

(i) *Scanner.* It is an electronic optical device that converts analog data such as maps drawn on a paper into the 'raster' format. A raster is a type of computerized picture logically made-up of a two-dimensional array of cells, like a spreadsheet. Aerial photographs and satellite imagery are common types of raster data used in GIS. Some

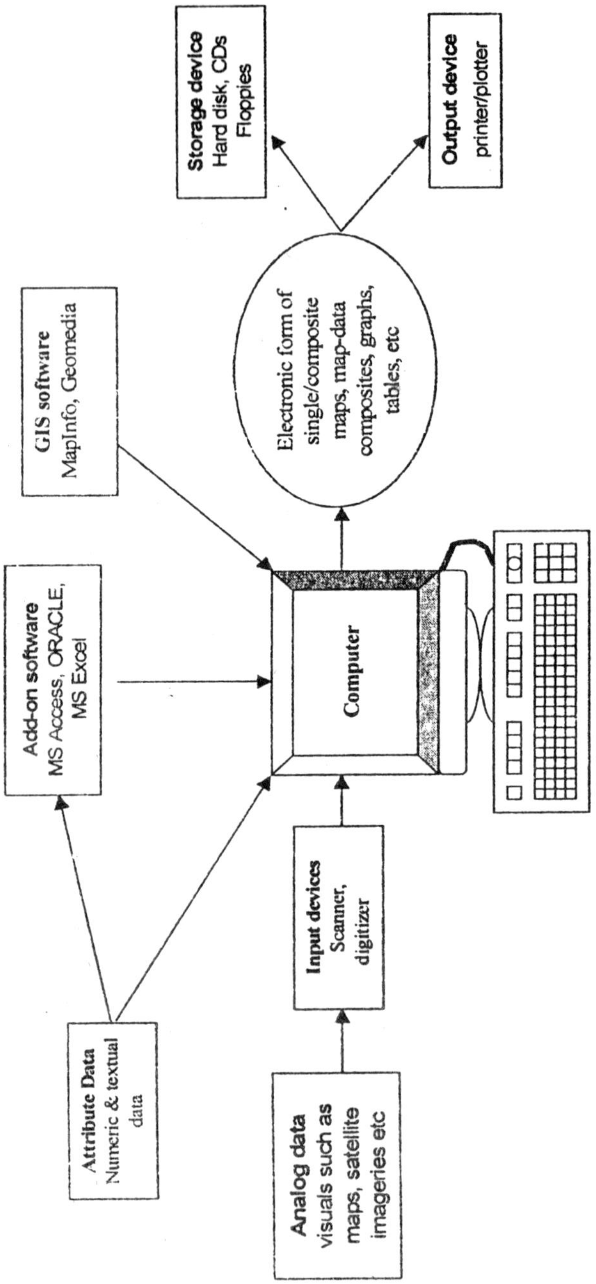

Fig. 1.1: Components of GIS

of the most familiar raster formats are TIF (Tagged Image File). BMP (Bitmap), and GIF (Graphics Interchange Format). A composite colour raster uses a colour table to map each raster cell value to a discrete display colour.

(ii) *Digitizer.* A device that enables converting any conventional paper map into such an electronic form that has specific position (in the form of coordinates) associated with each bit of the image. The digitizer consists of a table (or tablet) onto which a paper map is attached. The map is then traced by moving a hand-held, mouse-like device known as a cursor (also known as puck/transducer), across the surface. This results in the conversion of bits of the map into corresponding electronic version with coordinates associated with each bit.

A Vector is a co-ordinate based data structure commonly used to represent map features as point, line and polyline. Each object is represented with reference to x and y co-ordinates.

Some of the newer GIS software— *MapInfo* and *Geo Media* among others contain modules with which scanned images can be digitized onscreen without the digitizer hardware. As the onscreen digitizing provides features that are much more powerful and cost effective than the hardware-based digitizer, the later is becoming increasingly redundant.

(b) *Computer.* Computer forms the core hardware which stores and processes all information. Aside GIS software, a computer can also house statistical, graphical, and animation software which, depending on the user-friendliness of the GIS software, can be interfaced with the latter. The output can accordingly be enhanced in terms of quantity as well as quality.

(c) *Storage devices.* The maps and/or databases developed by the computer (with the help of GIS and other software) can be stored on digital media such as Compact Disc (CD), and Floppy Disc.

(d) *Output devices.* Printers and plotters constitute the output devices. Even a black-and-white dot-matrix printer can transfer a GIS map from its electronic form to a paper but to achieve distinct representation of various features in a map—with adequate tonal quality and contrast—colour inkjet/Laserjet printers of 1200 dpi or better resolution are needed.

3. Software

Some of the most popular GIS software are Arc Info, Arc GIS, MapInfo, Geo Media and TNT-MIPS. Each of these software offers different levels of functionality. It is in the interest of the GIS user to perform an assessment of the GIS requirements prior to committing purchase of the GIS software.

4. Users

Today, GIS is used by diverse professionals—from a cartographer to a commercial pizza dealer! Among the disciplines such as geology, hydrology and social sciences GIS has become as indispensable a tool as word processing and spread sheet software such as MS Excel.

To make the most of a GIS the user however need to be well versed in aspects such as map reading, database management, spatial analysis, computer cartography, computer science, programming, and basic geography.

Technical Elements of a Digital GIS

A GIS is built around a framework of five basic technical elements: (1) encoding (2) data input (3) data management (4) manipulative operations and (5) output products (Fig. 1.2).

Encoding

GIS mapping essentially deals with three distinct spatial attributes: points, lines and polygons. These spatial entities can be encoded by using two different types of position indexing systems: (1) grid-cell or raster coding and (2) polygon or vector coding. Grid-cell coding is conceptually a matrix system superimposed over the geography such that the attribute information can be collected by a systematic array of grid squares or cells. Normally, the information category most dominant for each cell is encoded.

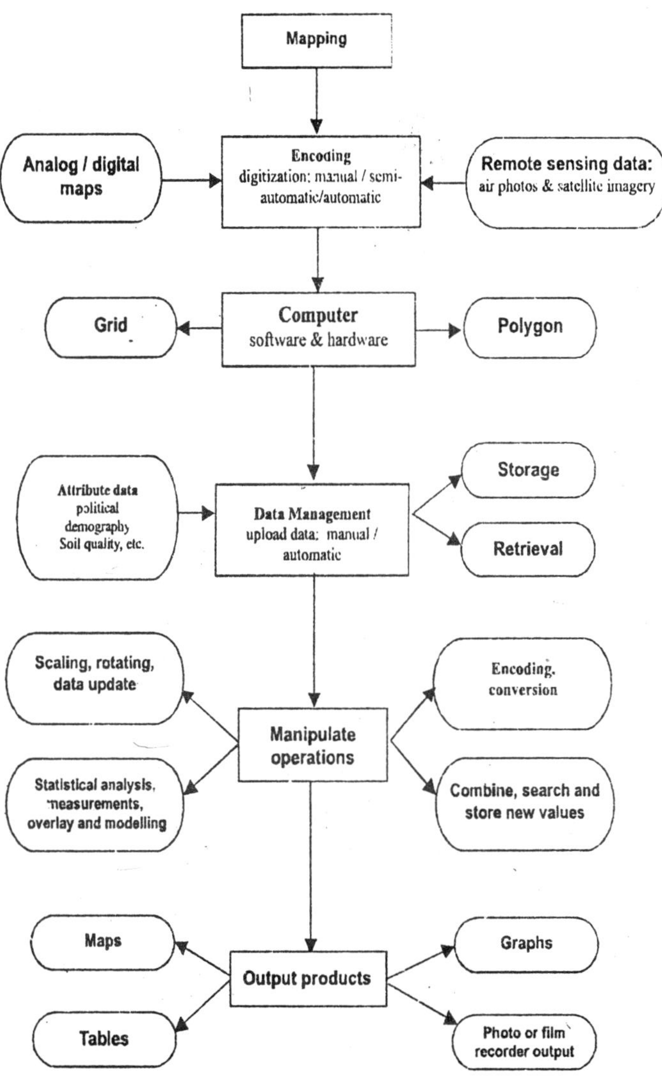

Fig. 1.2: Vatious steps involved in a typical GIS

As the cell size largely determines class accuracy, two methods have emerged to assist in preserving data integrity:

(a) decreasing cell size and (b) listing the relative amounts of each data type falling within a cell. Grid cells are functionally identical to the picture elements, or pixels, that compose a digital image.

In polygon coding, the perimeter of each unit area containing the desired attribute data is digitally encoded and stored. One type of polygon indexing is topological coding, whereby areas are formed by connecting polygons, and polygons are formed by connecting arcs. Polygon coding defines bound areas more accurately and requires less computer storage space than does the grid-coding structure.

Data Input

Analog information (e.g. maps on paper) is converted to the digital domain by the process called digitization. This forms an essential component of GIS. There are several methods for digitization: manual, semi-automatic and automatic.

Manual digitization involves tracing the analog maps using a specific hardware called table digitizer and other accessories such as puck/transducer. Semi-automatic digitization involves the use of a scanner, GIS software and the mouse of the PC. The analog maps after scanning (in the raster form) are registered using the GIS software. Once these scanned maps are registered, various topological features of these maps are redrawn and saved as several layers such as vegetation, settlements etc.,

The automatic digitizing involves hardware (a scanner) and the GIS software. The scanned maps or the raster images can be automatically digitized using some of the GIS software at one go. Later, the user would have to sieve the various digitized features, then select, cut and paste into various layers.

Data already in the digital form (e.g. satellite images) usually have to be reformated and scaled to match the geometry of the GIS reference map projection.

Data Management

Data management is extremely important for successful and

efficient operations of a GIS. Because of the large volume and the variety of data available these days, plus the wide range of potential applications. Data management consists of series of computer programmes to perform all data entry, storage-retrieval and maintenance tasks (Fig. 1.2).

Manipulative Operations

GIS are capable of performing two kinds of automated analysis, surface analysis and overlay analysis. Surface analysis applies to intra-variablle relationships that exist within one data plane. For example, soil categories can be grouped together, analyzed and labeled according to agricultural value. Most surface analysis produces new variables that can be applied to other surface or overlay analysis procedures.

Overlay analysis, as mentioned earlier, involves a set of distinctive map layers, the various features of the map stored in such a way that the user can view the layers of his choice.

Output Products

A GIS can retrieve and display data in graphic (as maps, bar charts, line graphs) or tabular form, or both. Most systems are capable of producing hardcopy charts, scatter diagrams, tables, and maps in various forms and sizes. In addition, all systems have a PC monitor on which graphic or tabular information for segments of a data base or multiple data base can be displayed. This represents interactive analysis because the retrieval and display of data are in near real time (Avery and Berlin 1992).

BASIC FUNCTIONS OF A GIS

User Interface

This is an important aspect of any software. It is necessary to choose software which provides a comprehensive set of capabilities, designed in such a way that the user can get used to the software in a very short time.

Some of the standard user interface of the GIS software are: wizard-like dialog boxes, short cut menus, status bar pop-ups, flexible toolbar positioning by dragging and the linked views of the map window and the browser window (Fig. 1.3 and 1.4).

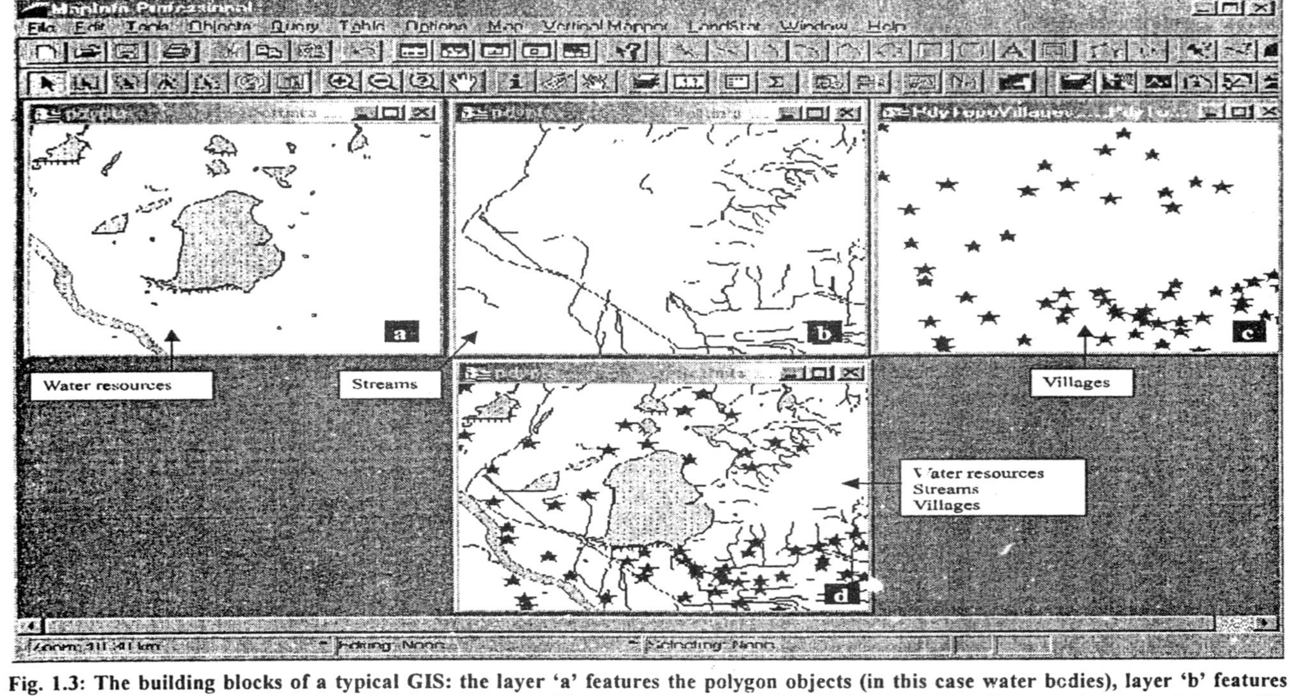

Fig. 1.3: The building blocks of a typical GIS: the layer 'a' features the polygon objects (in this case water bodies), layer 'b' features polylines (water streams) and the layer 'c' features point objects (villages). An overlay of the above three layers make a complete map (d).

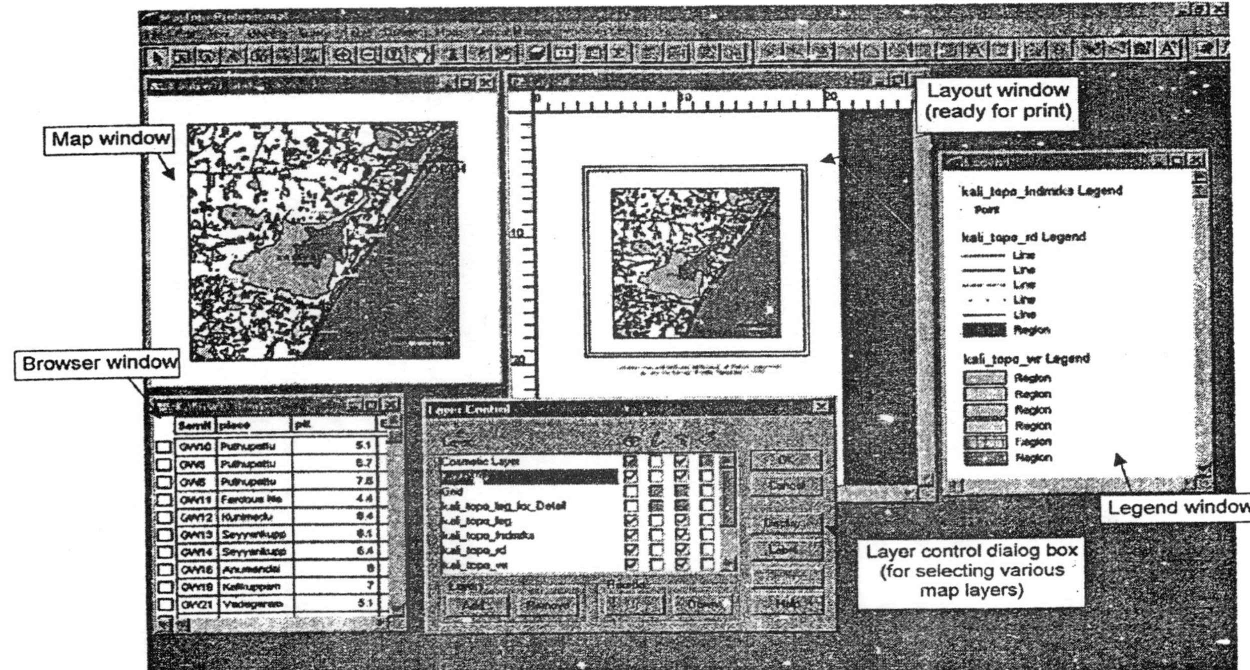

Figure 1.4: A typical interface of GIS software displaying various features

Mapping

As mentioned earlier, GIS enables dynamic mapping of the geography—earth and its features. Some of the features that make GIS dynamic are: layer control, legend display, zooming, and panning of the maps.

Layer control. This feature helps in overlaying several map layers and control them—which layer to be displayed and in which order.

Layer intelligence. When several map layers are opened, the GIS software orders the layers appropriately, i.e., points on top, boundaries underneath. This saves a lot of time by not requiring to reorder the maps while adding them.

Legend. Some software such as Arc View enables automatic display of map legends and in some software, such as MapInfo, it can be displayed by choosing the menu.

Zooming, panning and map auto scroll features add convenience by keeping the most important area visible.

Built-in-hot link capability enables one to display text files, images, other components of the project by clicking on the map object.

Data Analysis

The queries in GIS can be of geographic in nature, text based or through ODBC (Ole Data Base Connectivity). Queries in GIS can be expressed as avenue expressions or by SQL (Structured Query Language). Presently SQL based queries have gained much popularity among the users.

Desktop mapping tools enable the user to ask questions and get answers from the data. This helps in understanding the geographical relationships hidden in the database.

Information retrieval by clicking on the map. Usually the Info tools show the associated attributes of the map. The Button tools enable to select map objects or features by their proximity to other map objects.

SQL support: The query builder incorporates geographical

queries and functions. This includes geographic operators and functions in the relational SQL queries. This feature is highly useful in performing queries on multiple layers, for example one can select all the point objects that are present in the polygons of some other layer and vice versa.

Geographic operators: This feature provides the power of dynamic mapping. It enables the user to understand the geographical relationships within the database. Some of the most widely used geographic operators are: (i) completely within and completely contain (ii) have their center in, contain the centre of (iii) intersect (iv) are within the distance of.

Redistricting: This feature, also known as load balancing, allows the user to create new districts and realign existing districts, while doing calculations of the attached data for instant decision making. For example a manager can easily balance sales territories by the number of customers in the territories.

Find tool: A feature on the map or a record in the database can be easily searched using the find tool. This requires that the particular field is indexed, which is being searched.

Geographic Data Manipulation

Geographic data manipulation allows the user to create and edit geographic data and associate tabular information so that they can be analyzed more quickly and easily.

Polygon overlay capabilities with data aggregation and disaggregating makes it possible to combine, split, and erase map objects. This would make it easier to update the maps and their associated data .

Some of the popular tools for geographic object editing, which make digitizing and mapping easier are: overlaying of nodes at intersections, auto-tracing of polylines and polygons, convert polylines and polygons, snapping to the nearest nodes and auto-completing of polygons.

Buffer tool makes it easier to perform geographical queries and analysis. This feature is particularly helpful when performing multiple what-if scenarios. The data driven buffers, which use the underlying database information to determine buffer size, increases the flexibility in buffer creation and gives more accurate and/or more content rich means for visualizing the data. For example, for a layer of points that represent radio towers, one can create individual buffer zones based on each towers broadcast power.

Projection and Coordinate Systems: Support for various geographic projections is necessary when working with international data sets. Some of the standard features include: displaying of maps that are stored in numerous projections, supporting projection conversion, and supporting custom coordinate systems enables in digitizing, correctly overlaying maps, and to permanently change and save a map into a new projection.

Data Access

Large volumes of data are collected in different formats such as spreadsheets (MS Excel, Lotus 1-2-3), delimited ASCII files and relational databases (ORACLE, MS Access). Accessing these data in a fast and easy way with the desk top GIS would enhance the productivity and usability of a software.

Usually the spreadsheet data is accessed on the fly, while RDBMS (Relational Database Base Management system) can be accessed using the ODBC connectivity. Having a remote SQL read/write feature would not only save a lot of the time but makes it not necessary to store the intermediate copies of data locally.

Spatial Data (Map Files)

Compatibility: With a wide range of GIS software now catering to the requirements of GIS users, it is essential that the map data/files developed in one GIS software are readable/editable in other GIS software. Most of the GIS software support popular formats.

Universal translator: This feature enables the user to translate file formats of one software to the other required format.

Raster image support: This feature would enable to register raster images to the world coordinates and overlay these with other vector formatted data.

Spatial database support: Read/write support for SpatialWare, ORACLE, Autometrics, etc would allow to store geographic (map) data in RDBMS. This open architectural approach to client/server implementation of geographic queries eliminates re-geocoding of the data.

The compliance of database for point data (coordinates, appropriate spatial indices, etc) enables remote geographic queries to be performed on the server.

Spatial Statistics

The dominant feature of GIS technology is that spatial information is represented numerically, unlike in the case of analog paper maps. The analog nature of map sheets, manual analytic techniques are principally limited to qualitative processing. Digital representation on the other hand, has the potential for qualitative as well as quantitative processing.

The ever growing field of GIS has stimulated the development of spatial statistics, a discipline that seeks to characterize the geographic distribution or pattern of mapped data. Spatial statistics differs from traditional statistics by describing the more refined spatial variation in the data, rather than producing typical responses assumed to be uniformly distributed in space.

Map Algebra

Just as spatial statistics has been developed by extending concepts of conventional statistics, a spatial mathematics has evolved. It is similar to traditional algebra, in which primitive operations (add, subtract, exponentiation) are logically sequenced on variables to form equations, but in map algebra, entire maps represent variables. The logical sequence involves retrieval of one or more maps from the database, processing those data as specified by the user, creation of a new map containing the processing results, and the storage of the new map for subsequent processing.

The cyclical processing is similar to "evaluating nested parenthetical" in traditional algebra. Values for the "known" variables are first defined, then they are manipulated by performing the primitive operations on those numbers in the order prescribed by the equation. For example, in the equation A =

(B+C)/D the variables B and C are first defined and then added, with the sum stored as an intermediate solution. This intermediate value, in turn, is retrieved and divided by the variable D to derive the value of the unknown maps. The numbers contained in a solution map (in effect, solving for A) are a function of the input maps and the primitive operations performed. In a similar manner per cent change in any two maps can be calculated.

Visualization and Presentation

Visualization helps in understanding data through its visual representation. The end result of most of the GIS packages must have tools to make quality hardcopy or softcopy output.

A robust GIS software would provide multiple views of the data in the form of maps, charts, tabular views (rows and columns), and the ability to combine all these views (Fig. 1.4).

Maps-feature styles and labels: A good GIS software provides numerous style choices: pen styles (line styles, thickness, how end points meet), brush styles (fill patterns, including transparent fills), symbols (including custom symbols).

Maps-thematic representation: Thematic maps enable representation of data on a map with various colours, fill pattern, line styles, and symbols. Some of the thematic map types include: ranged maps (graduated by colour and size); individual (unique value); chart maps (pies and bars), dot density, and custom types); graduated symbol (the size of the symbol is proportionate to the data values of the points); bivariate or multivariate thematic analysis; and grid surface map (provides continuous colour gradation across the map).

These thematic map types enable the user to represent more than one variable in one map window. For example one can represent income by the size of the symbol and show education level by the symbol's colour. Also, it is possible to overlay more than one thematic map in a single map window. A polygon thematic map can be overlaid by the point based thematic map layer.

Charts and graphs: The GIS software also enables to create area, line, bar, pie and scatter plots with flexible display options (rotated, 3-D, etc) allowing to represent the data with more traditional business visualization tools without relying on other software applications.

Presentation. The final layout map with the WYSIWYG standards enable to make the final output ready for presentation.

Scalable Mapping Solutions

Scalable mapping solutions enable the user to leverage the power of geographic relationships regardless of the choice of application or implementation.

Scalable hardware requirements. This feature allows to adjust the hardware configuration based on the organization requirements. For the software which have provision with hardware lock can readily be upgraded with higher end RAM, HDD etc. However, in the case of floating/node locked software, which depend on the HDD identity or operating system, RAM or a combination of these, one would require to renew the licence.

Development environment: A desktop mapping development environment allows to: (a) extend the functionality of the base software (b) customize the user interface (c) automate repetitive processes (d) integrate with other applications.

GIS and Wetlands

GIS can play a crucial role in offsite identification of wetlands. GIS also helps in (i) bringing forth the hidden patterns in a data set (ii) interpolate the spatial data into a grids using several spatial models (iii) the grids can be used for preparing digital elevation models (DEM) that can be helpful in visualizing the drainage pattern (iv) perform query analysis using multiple data sets, and (v) as a powerful mapping media.

Remote Sensing and Wetlands

The inputs of remote sensing in the form of satellite imageries and air photos, aid in offsite identification of wetlands and in mapping/delineation of land use and land cover of the catchment and watershed.

Air Photos and Satellite Imagery

Wetlands can be interpreted using multi-band image types, such as multi-spectral scanners and hyper-spectral scanners, with at least one visible band and one near-infrared band. However, colour infrared photography enables better wetland interpretation because of the high level of contrast in image tone and colour between wetland and non wetland environments (Lillesand and Kiefer, 2002). Moist soil spectral reflectance patterns contrast more distinctively with less moist soils on colour infrared film than on panchromatic or normal colour films.

Several features of different satellite imageries, which enable identification and delineation of wetlands has been presented in Tables 1.1, 1.2, and 1.3.

Visual image interpretation can be used in a variety of ways to help monitor the quality, quantity, and geographic distribution of water resources (Lillesand and Kiefer, 2002).

The use of visual image interpretation coupled with selective field observations is an effective technique for mapping aquatic macrophytes (Lehman & lachavanne, 1999). More detailed information regarding total plant biomass or plant density can be achieved by utilizing quantitative techniques such as photographic density measurement techniques. Air photo interpretation has been used to monitor operations such as mechanical harvesting or chemical treatment of weeds.

Table—1.1: Landstat MSS bands and their environmental applications (Kudrat and Nag 1998)

Sensors	*Spectral range*	*Band name*	*Applications*
4	0.5-0.6	Green	Movement of sediment, laden water and delineation of shallow water
5	0.6-0.7	Red	Cultural features, discrimination of vegetation types
6	07-0.8	Near IR	Vegetation, boundary between land and water, landform
7	0.8-1.1	Near IR	Penetration of atmospheric haze, vegetation, boundary between land and water and landform

Table—1.2: Spectral bands of Landsat Thematic Mapper (TM) (Kudrat and Nag 1998)

Band	*Spectral range*	*Applications*
TM1	0.45-0.52	Designed for water body penetration useful in coastal bathymetry, differentiation of soil from vegetation, and deciduous from coniferous flora
TM2	0.52-0.60	Measurement of visible green reflectance peak of vegetation for vigor assessment
TM3	0.63-0.69	Chlorophyll absorption band, important for vegetation discrimination
TM4	0.76-0.90	Determination of biomass content and surface water mapping
TM5	1.55-1.75	Indicative of vegetation moisture content and soil moisture, also useful for separating snow from clouds
TM6	10.40-12.50	Thermal channels useful in vegetation stress analysis, soil moisture discrimination and thermal mapping
TM7	20.08-2.35	Discrimination of rock types and for hydro-thermal mapping

Table—1.3: IRS spectral bands and their applications (Kudrat and Nag 1998)

Band	*Spectral range*	*Applications*
1	0.45-0.52	Coastal environment studies, soil/vegetation differentiation, Coniferous/deciduous vegetation discrimination
2	0.52-0.59	Vegetation vigour, rock/soil discrimination, turbidity and bathymetry in shallow waters
3	0.62-0.68	Strong chlorophyll absorption leading to discrimination of plant species
4	0.77-0.86	Delineation of water features, landform geomorphic studies

Application of GIS and Remote Sensing in Wetland Management

We present below a few illustrative examples where GIS and remote sensing have been successfully applied in wetland management.

1. Impact of Human Interference on Kaliveli Wetland

Kaliveli, about 130 Km south of Chennai and close to Pondicherry is one of the largest wetlands of Asia. It is a very important wintering ground for migratory birds and has been identified as a heritage cite by the International Union for the Conservation of Nature (Abbasi 1997). In order to develop a system which could continuously assess the impact of human activities on Kaliveli wetland, Chari and Abbasi (2001) have developed a GIS by integrating topo sheet maps, satellite imageries, and ground truth studies with the help of software tool MapInfo Professional 5.5. The resulting GI system not only helps in identifying impacts but also throws up suggestions on how to manage the wetland in an integrated, sustainable manner.

2. Monitoring Small Dams in Semi-arid Regions Using Remote Sensing and GIS

The analysis of data from high spatial resolution satellite-mounted infrared sensors has the potential to monitor the water stored by small dams in semi-arid areas (Finch, 1997). The author has described a simple method of analysis and particular attention is given to selecting an appropriate threshold level to discriminate water from non-water land cover. The use of a GIS allowed further discrimination of the areas of water from other areas, such as deep shadow, which have been classified incorrectly and facilitates the automatic calculation of dam capacities. The success of the technique was demonstrated on data from two regions in Botswana.

3. Assessing the Vulnerability to Soil Erosion of the Ukai Dam Catchments Using Remote Sensing and GIS

The investigation of basins for planning soil conservation requires a selective approach to identify smaller hydrological units, which would be suitable for more efficient and targeted conservation management programmes. One criterion, generally used to determine the vulnerability of catchments to erosion, is the sediment yield of a basin. In India, sediment yield data are generally not collected for smaller sub-catchments and it becomes difficult to identify the most vulnerable areas for erosion that can be treated on a priority basis. An index-based approach, based on the surface factors mainly responsible for soil erosion, is suggested in a study by Jain and Goel (2002). These factors include soil type, vegetation, slope and various catchment properties such as drainage density, form factor, etc.

The method adopted by Jain and Goel (2002) has been illustrated with a case study of sub-catchments immediately upstream of the Ukai Reservoir located on the River Tapi in Gujarat State, India. The area is divided into 16 watersheds and different soil, vegetation, topography and morphology-related parameters are estimated separately for each watershed. Satellite data are used to evaluate the soil and vegetation indices, while a GIS system is used to evaluate the topography and morphology-related indices. The integrated effect of all the parameters is evaluated to find different areas vulnerable to soil erosion. Two watersheds were identified as being most susceptible to soil erosion. Based on the integrated index, a priority rating of the watersheds for soil conservation planning is recommended.

4. Geographic Information Systems (GIS)-based Spatially Distributed Model for Runoff Routing

Olivera and Maidment (1999) have proposed a method for routing spatially distributed excess precipitation over a watershed to produce runoff at its outlet. The land surface is represented by a (raster) digital elevation model from which the stream network is derived. A routing response function is defined for each digital elevation model cell so that water movement from cell to cell can be convolved to give a response function along a flow path and responses from all cells can be summed to give the outlet hydrograph.

5. Changes in the Water Quality Indicated by Submerged Macrophytes Using GIS

Geographic Information Systems (GIS) are useful for mapping and storing information on submerged vegetation allowing easy interrogation, updating and plotting of spatial information at various scales, and providing a reference for future comparisons.

Lehmann and Lachavanne (1999) have studied the distribution of submerged macrophytes along 20 km of lake shore (Lake Geneva, Switzerland) was compared between the years 1972, 1984 and 1995. Lake Geneva underwent rapid eutrophication until 1980, followed by reversal that is still in progress. *Potamogeton pectinatus, P. perfoliatus, P. lucens and Elodea Canadensis* showed no significant changes in their distribution, with the two former species dominant throughout. *Chara sp.*, and to a lesser degree *Myriophyllum spicatum,* decreased in abundance between 1972 and 1984 but had increased again by 1995. The abundance of *P. Pusillus* increased regularly, while *Zannichellia palustris* and *P. crispus* almost disappeared from the study area. *Elodea nuttallii* was observed for the first time in Lake Geneva in 1995.

Two methods of bioindication of water quality by macrophytes are compared. The macrophyte index proposed by Melzer (1988) is based on nutrient load, whereas the saprobic index proposed by Sladecek (1973) measures organic pollution. The saprobic index is sensitive to small changes in species composition and abundance, and also reflects better the changes in eutrophication. It may therefore be a better bioindicator.

6. Integrated Geographical Assessment of Environmental Condition in Water Catchments

Water catchments are functional geographical areas that integrate a variety of environmental processes and human impacts on landscapes. Integrated assessments recognize this interdependence of resources and components making up water catchments and are vital for viable long-term natural resources management. The work by Aspinall and Pearson (2000) couples eco-hydrological modelling with remote sensing, landscape ecological analyses and GIS to develop a series of indicators of water catchment health as part of a geographical audit of environmental health and change at regional scales.

Indicators are simple measures that represent key components of the system and have meaning beyond the attributes that are directly measured. A suite of indicators, many capable of measurement from remote sensing data sources, are described that represent state (condition) and trend (changes across space and time) and focus on the physical, biological and chemical properties of water catchments, as well as their ecological function (stability, resilience, and sensitivity). Models implemented in GIS allow indicators to be combined within water catchments by setting them within a specific geographic context and integrating the descriptions of environmental variability across the geographic area. This spatial integration is necessary to place individual, site-specific indicators within a broader geographic context; the models allow this context to reflect the ecological and hydrological functioning of the water catchment. Scale and other geographic effects associated with integration are managed using an approach that partitions the landscape into a hierarchical series of nested functional units.

Methods from image analysis, landscape ecological analysis, spatial interpolation, and numerical process modelling are integrated within a GIS (Arc View) to provide a single environment within which to conduct the study. Results are described from the catchment of the upper Yellowstone River in the Rocky Mountains, USA, an area of about 14000 km^2. The river source is in Yellowstone National Park. The catchment is subject to a number of land-use issues notably those associated with changing patterns and types of land use including forestry, irrigated agriculture, range management, wildfire, mining, summer and winter recreation, and residential development, which are associated with a number of land-use conflicts and impacts.

7. Application of GIS in Fisheries Science

Introduction of GIS in fisheries science has been relatively recent compared to other disciplines. This is due, in part, to the inherent complexity of the environment where fishes life cycles take place and to the complexity of fisheries dynamics as well. Geographic information systems have been developed mainly for terrestrial landscapes and many of their algorithms are not suitable for marine ecosystems. The principal challenges in the implementation of geographic information systems to fisheries research have to do with the three dimensional space where fish live, temporal variability and fuzziness of the data sets. There are still few studies in fisheries science which incorporate GIS in their research. In a research article Malavear (2002) has reviewed some of the areas in fisheries research which are currently using geographic information systems, the positive outcomes of the integration of GIS in fisheries science, enumerate some methodological problems found and how these studies deal with them and finally outline some potential areas for further research.

8. Applying GIS and Landscape Ecological Principles to Evaluate Land Conservation Alternatives

Lathrop and Bognar (1998) have discussed the conflict between the conservation of the rich biological diversity of the existing forested landscapes of eastern United States and a continued expansion of suburban/exurban development. The authors have used geographic information systems (GIS)-based assessment and landscape ecological principles to assess the environmental sensitivity of Sterling Forest lands, a 7245 ha tract of land on the New York-New Jersey border, and prioritized lands for conservation protection. This GIS assessment served as the basis of subsequent negotiations of a compromise conservation-development plan by a coalition of land conservation trusts and the land owner/developer. Sterling Forest represents a useful case study of the application of GIS technology by the non-profit environmental groups is successfully undertaking an independent analysis of a regionally important land use issue.

9. Application of Remote Sensing and GIS for Surveillance of Mosquito Vector Habitats and Risk Assessment

Dale et al (1998) have successfully used remote sensing to achieve multiple objectives focusing on mosquito management. The authors have demonstrated how Geographic Information Systems, combined with remote sensing analysis, have the potential to assist in minimizing disease risk. Examples are used from subtropical Queensland, Australia, where the salt marsh mosquito, *Aedes vigilax*, and the freshwater species, *Culex annulirostris*, are vectors of human arbovirus diseases such as Ross River and Barmah Forest virus disease. *Culex annulirostris* is also implicated in the transmission of Japanese Encephalitis. Mapping the breeding habitats of the species facilitates assessment of the risk of contracting the diseases and also assists in control of the vectors. First, it considers a simple risk model that is applied to data for the city of Brisbane is southeast Queensland. This is then linked to computer-aided analysis of remotely sensed data to map potential ephemeral freshwater breeding sites of *Cx. annulirostris*. This has the potential to guide control at critical times, for example after heavy summer rainfall or when there is an outbreak of Ross River virus disease. Second, the use of colour infrared aerial photography is used to identify the specific parts of the salt marsh in which larvae and eggs of *Ae. vigilax* are found. The authors also explore novel ways to map the detailed pattern of water under mangrove forest canopy to identify where mosquitoes are breeding and as an aid to planning modification. For each we discuss the limitations and advantages and the possibilities for combining methods and/or using a single method for multiple objectives.

10. Visualization Techniques for Incorporation in Forest Planning Geographic Information Systems

Visual representations are increasingly used to communicate the impacts of environmental changes. Geographic information systems (GISs) are becoming common sources of the spatially organized data needed to create valid and defensible visualizations. However, these data lack the detail and richness needed to create the realistic imagery felt to be critical for public review. The paper by Orland (1994) describes exploratory studies in the application of techniques drawn from remote sensing and applied to ground-level photographic images of sensitive locations to achieve realistic images with demonstrable relationships to an underlying GIS. Digital filtering and image sampling processes have been used to simulate the visual consequences of forest pests, of timber management activities, of forest wildfires, and of recovery from all of these impacts. The resulting images have been used to communicate expected outcomes to participants in policy-development settings, and to initiate the development of public perception models relating impacts to public preferences. Although integration of these techniques with GIS systems has not yet been achieved, the necessary development steps are outlined by the authors.

REFERENCES

Abbasi S. A., 1997. *Wetlands of India and Ecology and Threats: the Ecology and the Exploitations of Typical South Indian Wetlands*, Vol I, Discovery Publications House, New Delhi, p: 149.

Aronoff S., 1989. Geographic Information Systems: A Management Perspective, WDL Publications, Ottawa, Canada.

Aspinall R., and Pearson D., 2000. *Integrated Geographical Assessment of Environmental Condition in Water Catchments: Linking Landscape Ecology, Environmental Modelling and GIS*, Journal of Environmental Management, 59, pp. 299-319.

Avery T. E., Berline G. L., 1992. *Fundamentals of Remote Sensing and Air-photo Interpretation*, Mac Millan Publishing Company, New York.

Burrough P.A., 1998. *Principles of Geographic Information Systems for Natural Resource Assessment*, Oxford University Press, London, p. 194.

Burrough P.A., and McDonnell R. A., 1998. *Principles of Geographic Information Systems*, Oxford University Press, London, p. 333.

Carter J. R., 1989. *On Defining the Geographic Information Systems*, In Ripple W. J.; (ed.), Fundamentals of Geographical Information Systems: A compendium, ASPRS/ACSM, Falls Church, Va., pp. 3-7.

Chari K. B., and Abbasi S. A., 2001. *Assessing Land Use/land Cover of Kaliveli Watershed Using Remote Sensing and GIS: Implications for Conservation*, In Spatial Information Technology: Remote Sensing and GIS, Muralikrishna I.V. (ed.), Vol II, BS Publications, Hyderabad, pp. 331-336.

Cowen D. J., 1988. *GIS Versus CAD Versus DBMS: What are the Differences?*, Photogrammetric Engineering and Remote Sensing, pp. 54: 1551-4.

Dale P. E. R., Ritchie S. A., Territo B. M., Morris C. D., Muhar A., and Kay B. H., 1998. *An Overview of Remote Sensing and GIS for Surveillance of Mosquito Vector Habitats and Risk Assessment*, Journal of Vector Ecology 23(1): 54-61.

Department of Environment (DoE), 1987. *Handling Geographic Information System*, HMSO, London.

Finch J. W., 1997. *Monitoring Small Dams in Semi-arid Regions Using Remote Sensing and GIS*, Journal of Hydrology, pp. 195: 335-351.

Jain S. K., Goel M. K., 2002. *Assessing the Vulnerability to Soil Erosion of the Ukai Dam Catchments Using Remote Sensing and GIS*, Hydrological Sciences Journal-Journal Des Sciences Hydrologiques, pp. 47(1): 31-40.

Kudrat M., and Nag P., 1998. *Digital Remote Sensing*, Concept Publishing Company, New Delhi, p. 320.

Lathrop R. G., and Bognar J. A., 1998. *Applying GIS and Landscape Ecological Principles to Evaluate Land Conservation Alternatives*, Landscape and Urban Planning, pp. (41): 27-41.

Lehmann A., and Lachavanne J. B., 1999. *Changes in the Water Quality of Lake Geneva Indicated by Submerged Macrophytes*, Freshwater Biology, pp. 42:457-66.

Malavear M. Y., 2002. *The Applications of GIS to Fisheries Science: Recent Trends, Methodological Problems and Challenges*, web article.

Martin D., 1996. *An Assessment of Surface and Zonal Models of Population*, International Journal of Geographical Information Systems, pp. 10: 973-89.

Melzer A., 1988. *Die Gewasserbeurteleilung Bayerischer Seen mit Hilfe Makrophytischer Wasserpflanzen Gefahrdung und Schutz Von Gewassern*, Ulmer V. E., (Ed.), Universitat Hohenheim, Stuttgart, pp. 106-116.

Olivera F., and Maidment D., 1999. *Geographic Information Systems (GIS)-based Spatially Distributed Model for Runoff routing*, Water Res rces Research, pp. 35(4)1155-64.

Orland B., 1994. *Visualization Techniques for Incorporation in Forest Planning Geographic Information Systems*, Landscape and Urban Planning, pp. 30: 83-97.

Ozemoy V. M., Smith D. R., and Sicherman A., 1981. *Evaluating Computerized Geographic Information Systems Using Decision Analysis*, Interfaces, pp. 92-98.

Parker H. D., 1988. The *Unique Qualities of a Geographic Information System:A Commentary*, Photogrammetric Engineering and Remote Sensing, pp. 54: 1547-9.

Smith T. R., Menon S., Starr J. L., and Estes J. E., 1987. *Requirements and Principles for the Implementation and Construction of Large-scale Geographic Information Systems*, International Journal of Geographic Information Systems, pp. 1:13-31.

Sladecek V., 1973. *System of Water Quality from the Biological Point of View*, Archiv fur Hydrobiologie, Bieheft Ergenbnisse der Limnologie, Heft, pp. 7:1-218

Lillesand T.M., and Kiefer R.W., 2002. *Remote Sensing and Image Interpretation*, John Wiley and Sons, New York, p. 724.

2

Wetlands

An Overview

Abstract

Wetlands have always been important but in recent years their value has been increased in direct proportion to the threats wetlands are facing from population and pollution.

This paper presents a state-of-the-art review on the classification, importance, and status of wetlands, with particular emphasis on Indian wetlands. The role of the two most modern and powerful tools—remote sensing and GIS—in wetland management is highlighted. Also, some of the principles of eco-restoration and management of wetlands, which are widely applied, have been presented.

Key words: wetlands, GIS, remote sensing, wetland delineation, eco-restoration, management

Introduction

Since time immemorial wetlands have served as cradles of civilizations, the focal points around which great cultures took roots and flourished (Abbasi, 1997). Human beings, living in the flood plains of major rivers used the annual cycles of the floods and the nutrients deposited on the plains for their advantage.

Wetlands are among the most productive and biologically rich ecosystems on earth (Richardson, 1995). Wetlands are also amongst the most endangered; the feverish pace of urbanization and industrialization during recent decades has resulted in the worldwide destruction of wetlands.

In this paper, an introduction to wetlands with special reference to GIS and remote sensing has been presented. Eco-restoration and management of wetlands has also been discussed.

What are Wetlands?

Wetlands, as the name suggest, refer to such a landscape which is saturated with water or covered by water either perennially or for a major part of the year. There are numerous definitions of wetlands but all essentially agree to the above criterion (Ambasht et al, 1990; Mitsch and Gosselink, 1993).

The Ramsar convention (1971) of the IUCN has been an initial step to define wetlands and to mobilize national and international government action to fight their rapid loss. As per the definition given at Ramsar convention (Scott,1989) wetlands are areas of marsh, fenn, peat lands of water, whether natural or artificial, permanent or temporary; with water that is static or flowing; fresh, brackish or saline, including areas of marine water the depth of which at the low tide does not exceed six meters.

According to Cowardin et al. (1979), wetlands are lands transitional between terrestrial and aquatic systems where the water table is at or near the surface, or the land is covered by shallow water.

A detailed classification of wetlands based on Dugan (1990) has been presented in Fig. 2.1.

In India, Kaul et al. (1978) and Trisal and Zutshi (1985) have made serious attempts at classifying wetlands on the basis of vegetation. They have classified wetlands into herbaceous and forested types (Fig. 2.2).

Significance of Wetlands

Wetlands are one of the most productive ecosystems of the world; their primary productivity is known to exceed that of grasslands, cultivated lands and even tropical rain forests (Richardson, 1995) (Fig. 2.3).

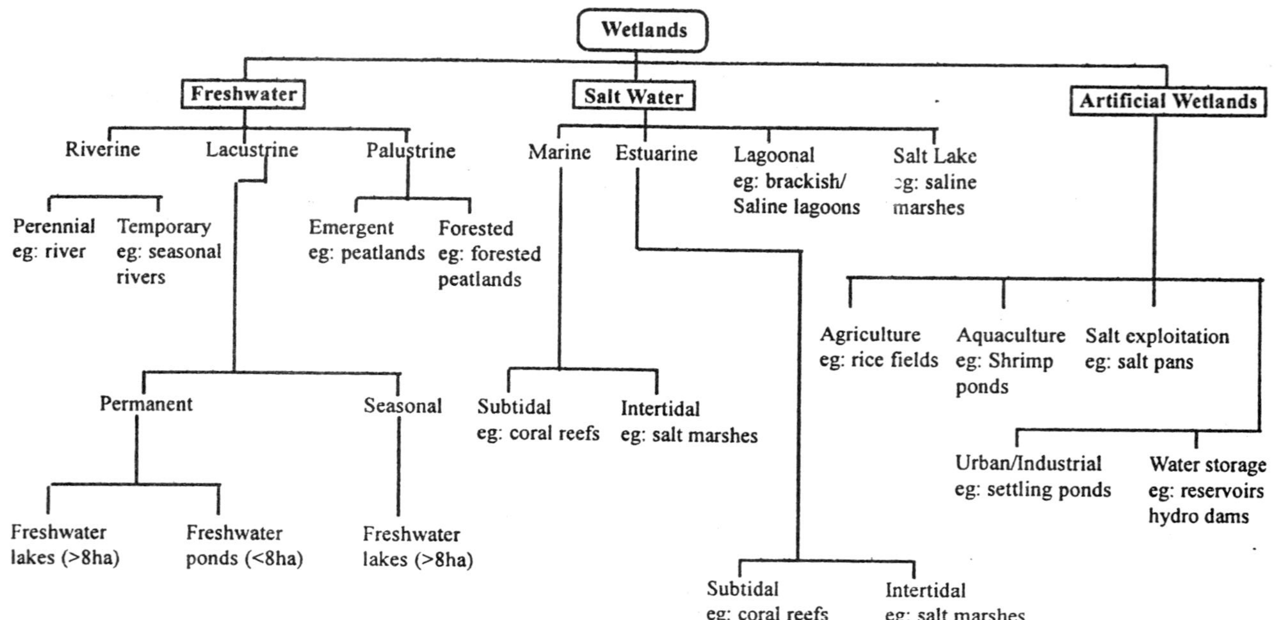

Fig. 2.1: Classification of wetlands (Dugan, 1990)

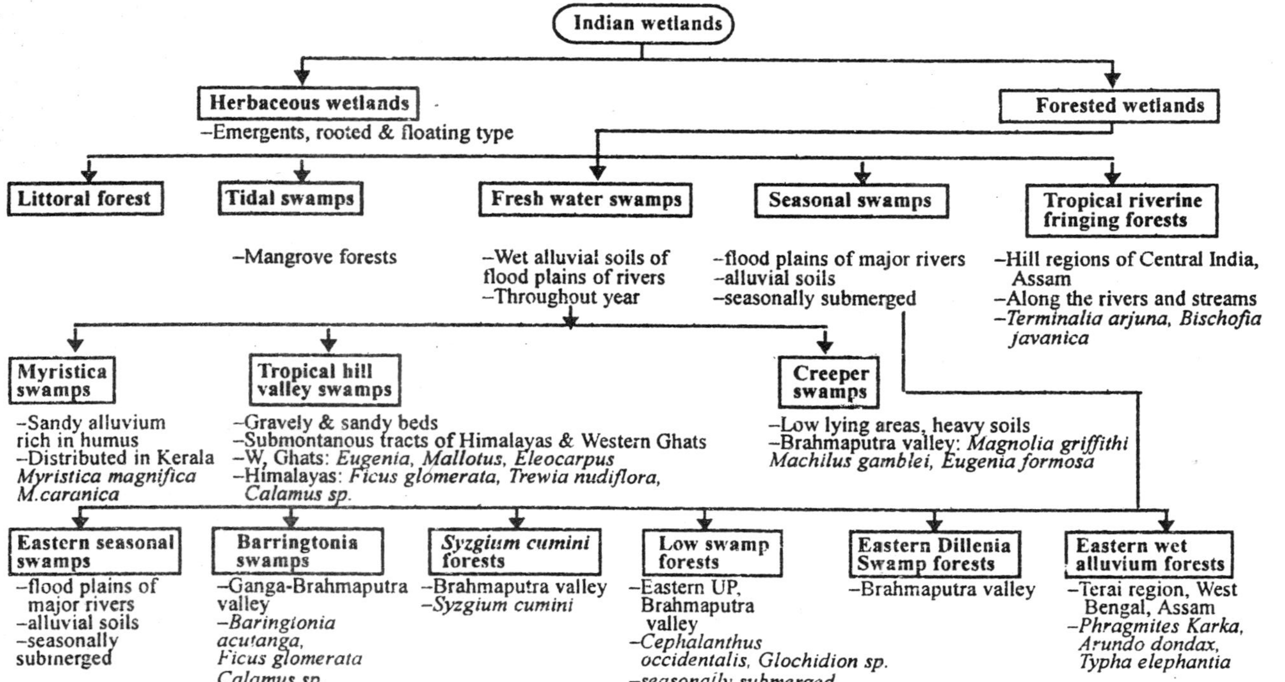

Fig. 2.2: Classification of Indian wetlands based on vegetation types (Singh 1996)

Wetlands, apart from being the most productive ecosystems, are also known for several hydrological and ecological functions which are summarized in the Table 2.1.

Identification of Wetlands

The accurate identification of wetlands is essential for the conservation of wetlands. They differ widely in character, even in a local region, due to differences in soils, topography, hydrology, water chemistry, vegetation, and other factors. Contributing to the confusion on what constitutes a wetland is the fact that different agencies have jurisdiction over wetland areas and different definitions delimiting them. Because wetlands can vary so significantly within a given type, no universally recognized wetland definition/criterion exist.

In the following sections, a brief description of the various criteria for identification of wetlands has been presented.

Ramsar Convention

The criterion adopted by the contracting parties of the Ramsar convention, held in July 1990, for identifying wetlands has been presented below:

(a) ***Criteria for Representative or Unique Wetlands***

(i) It is a particularly good representative example of a natural or near-natural wetland, characteristic of the appropriate biogeographic region.

(ii) It is a particularly good representative example of a natural or near natural wetland, common to more than one biogeographic region.

(iii) It is a particularly good representative example of a wetland, which plays a substantial hydrological, biological or ecological role in the natural functioning of a major river basin or coastal system, especially where it is located in a trans-border position.

(iv) It is an example of a specific type of wetland, rare or unusual in the appropriate biogeographical region.

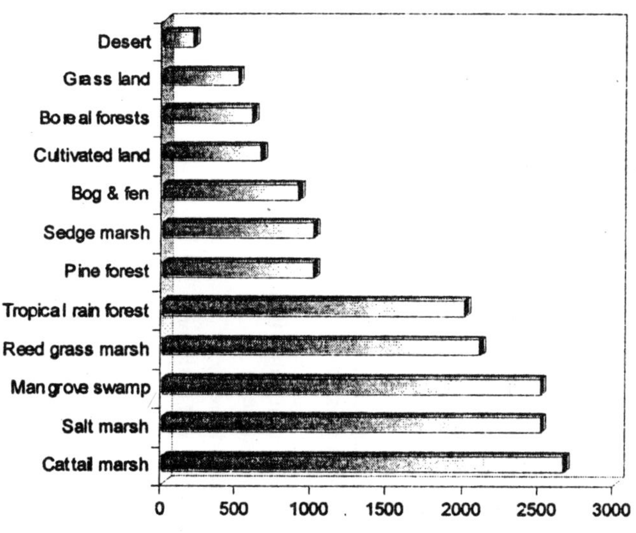

Fig. 2.3: Comparison of primary productivity of various ecosystems

(b) General Criteria Based on Plants and Animals

(i) It supports an appreciable assemblage of rare, vulnerable or endangered species or subspecies of plants or animals or an appreciable number of individuals of any one or more of these species.

(ii) It is of special value for maintaining the genetic and ecological diversity of a region because of the quality and peculiarities of its flora and fauna.

(iii) It is of special value as the habitat of plants or animals at a critical stage of their biological cycle.

(iv) It is of special value for its endemic plant or animal species or communities.

(c) Specific Criteria Based on Waterfowl

(i) It regularly supports 20,000 waterfowl: a substantial number of individuals from particular groups of waterfowl indicative of wetland values, productivity, or diversity.

Table—2.1 Various functions of wetlands (Keddy, 2000)

Regulation functions	*Carrier Functions*	*Production Functions*	*Information Functions*
• Regulation of the local and global energy balance • Regulation of the chemical composition of the atmosphere • Regulation of the chemical composition of the oceans • Regulation of the local and global climate (including Hydrological cycle) • Regulation of runoff and flood prevention (watershed protection) • Water catchment and groundwater recharge • Prevention of soil erosion and sediment control • Formation of topsoil and maintenance of soil fertility • Fixation of solar energy and biomass production • Storage and recycling of organic matter • Storage and recycling of nutrients • Storage and recycling of human wastes • Regulation of biological control mechanisms • Maintenance of migration and nursery habitats • Maintenance of biological (and genetic) diversity.	• Human habitation and (indigenous) settlements • Cultivation (crop growing, animal husbandry, aquaculture) • Energy conversion • Recreation and tourism • Nature protection	• Oxygen • Water (for drinking, irrigation, industry etc.) • Food and nutritious drinks • Genetic resources • Medicinal resources • Raw materials for clothing and household fabrics • Raw materials for building, construction and industrial use • Biochemical (other than fuel and medicines) • Fuel and energy • Fodder and fertilizer • Ornamental resources	• Aesthetic information • Spiritual and religious information • Historic information (heritage value) • Cultural and artistic inspiration • Scientific and educational information

(ii) Where data on populations are available, it regularly supports 1% of the individuals in a population of one species or subspecies of waterfowl.

Criteria Based on the US ACOE Manual (1987)

As per the criteria based on the 1987 manual of US ACOE, the indicators of the presence of a wetland are hydrophytic vegetation (plant life growing in water, soil, or on a substrate that is periodically deficient in oxygen due to excess water), presence of water, and hydric soils (soils saturated, flooded, or ponded, long enough during the growing season to develop anaerobic conditions in the upper profile) (Watersheds).

Offsite Identification of Wetlands

As mentioned earlier, in Chapter 1, wetlands can be identified offsite with the help of GIS. However, on-site verification is necessary to establish the existence, size, shape, and type of wetlands. Some resources for offsite identification of wetlands has been presented below:

Survey of India (SOI) Topographic Maps

The SOI toposheets of the scale 1:50,000 portray surface features such as rivers, lakes, canals, inundated/waterlogged areas, marshes, and other vegetation cover types. However, rice fields and other wetland types that are manipulated by man do not feature in them. These maps though are of large scale serve as a good source for generating base maps.

The village maps, at the scale of 1:10000, are useful in identifying the village boundaries, major water resources and canals. However, these maps do not feature latitude and longitude information nor map scale. Thus, one has to rely on the prominent map features such as intersecting roads or canals for digitizing. These maps, being imprecise, do not fit exactly over the SOI toposheets. However, for developing micro level GIS (geographic information system) these maps would be of immense value as they feature village boundaries and vegetation which are not found in the SOI toposheets of the scale 1:50,000.

Satellite Imageries and Air-photos

Today we have a wide array of satellite imageries and air-photos which can help in the identification and real time monitoring of wetlands. Some of the widely used imagery systems for wetland mapping and identification are Landsat Thematic Mapper, SPOT, and IRS.

The Indian space programme has launched six satellites for remote sensing-IRS-1A, IRS-1B, IRS-P2, IRS-1C, IRS-D and IRS-E. National Remote Sensing Agency (NRSA), Hyderabad, is concerned with the dissemination of IRS and other satellite imageries in India. Today, regional remote sensing centers, state remote sensing centers; several national and regional institutions; and universities have facilities for handling remote sensing data.

Directory of Wetlands (MoEF, 1990)

Ministry of Environment and Forests (MoEF), Government of India, has published a directory of wetlands (1990) based on the survey carried out during 1972. However, the survey is not comprehensive and many inland wetlands and most of the coastal wetlands have not been included in the compilation.

A Nation-wide Wetland Mapping Project

A Nation-wide Wetland Mapping Project, sponsored by the Ministry of Environment and Forests (India), was carried out by Space Applications Centre (ISRO), Ahmedabad, India (MoEF Report). Wetland delineation and mapping has been done using IRS LISS IIII data of 1992/1993 (pre-monsoon and post-monsoon seasons) using visual analysis techniques. For the states where cloud free satellite data of 1992/93 time-frame was not available, data of 1991/92 or earlier years has been used. Mapping/ monitoring for 21 notified wetlands and their catchments has also been done on 1:50 000 scale using IRS LISS II data.

Results of the wetland inventory have been summarised state-wise. Results are discussed based on type of the wetlands and also district-wise for each state. As a part of the project, database has also been prepared for facilitating monitoring at a later date.

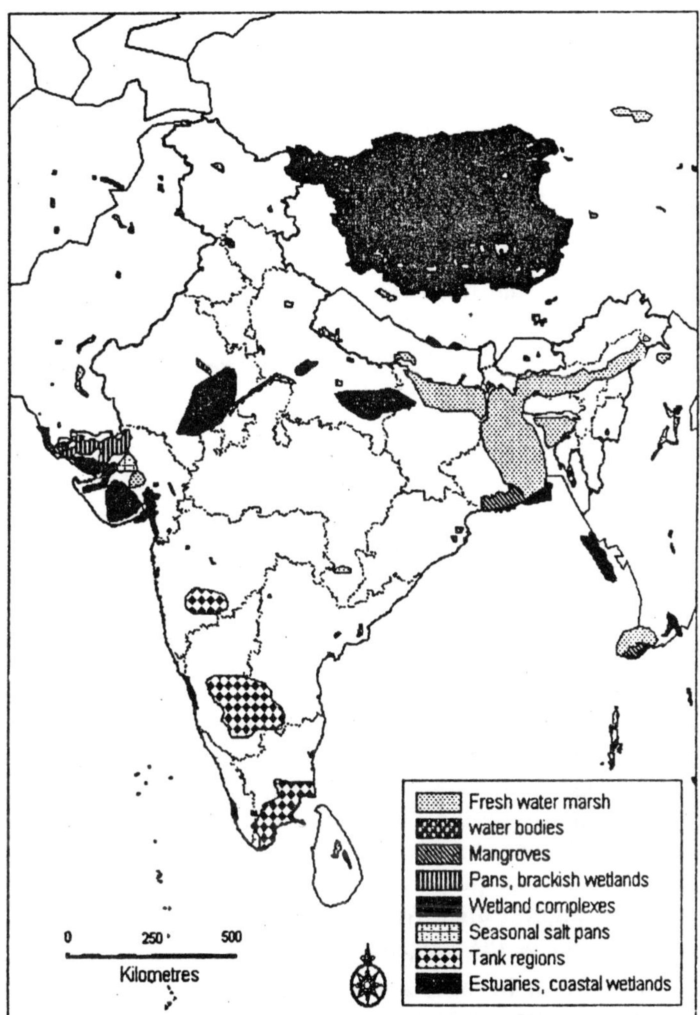

Fig. 2.4: Location of various types of wetlands of India, redrawn from the map of World Conservation Monitoring Centre, which in turn was prepared on the basis of the mapped information contained on the 'A Directory of Asian Wetlands' by Derek A. Scott (1989).

Wetlands of India

India has a rich variety of wetland habitats (Fig. 2.4). The total wetland area in the country-excluding paddy rice, rivers and canals-has been estimated to be about 7.6 Mha, out of which 3.6 Mha are inland and 4 Mha are coastal (MoEF Report).

Wetlands of India can be differentiated region-wise into eight categories (Scott, 1989): the reservoirs of the Deccan Plateau in the south, together with the lagoons and the other wetlands of the southern west coast; the vast saline expanses of Rajasthan, Gujarat and the gulf of Kutch; freshwater lakes and reservoirs from Gujarat eastwards through Rajasthan (Keoladeo Ghana National Park) and Madhya Pradesh; the delta wetlands and lagoons of India's east coast (Chilka Lake); the freshwater marshes of the Gangetic Plain; the floodplain of the Brahmaputra; the marshes and swamps in the hills of north-east India and the Himalayan foothills; the lakes and rivers of the montane region of Kashmir and Ladakh; and the mangroves and other wetlands of the island arcs of the Andamans and Nicobars.

The Present Status of Indian Wetlands

The wetlands across the world are suffering from 'ecological coma' the recovery from which is virtually impossible (Llamas,1994), and Indian wetlands are no exception to this. Wetland habitats of India have been destroyed by draining and land filling, industrial pollution, and other demographic pressures. Though figures of the overall loss of wetlands across India are not available, a study by the Wildlife Institute of India reveals that some 70-80 percent of individual fresh water marshes and lakes in the Gangetic flood plains have been lost in the last 50 years. The country's mangrove areas have been almost halved from 700,000 hectares in 1987 to 453,000 hectares in 1995.

Since India became a contracting party to the Ramsar Convention in 1981, only six wetlands in the country have been designated as Ramsar Sites. Recently, MoEF has proposed another 10 new wetlands, covering an area of 1.1 million ha, to be listed as Ramsar sites in 2003 (Ramsar Wetland Bureau) (Fig.2. 5). The Ramsar listing of the Indian wetlands is just a fraction of the wealth of aquatic ecosystem types in the country. There is, therefore, a need for identifying additional wetlands for listing under the Ramsar Convention and thus conserve them before we lose them.

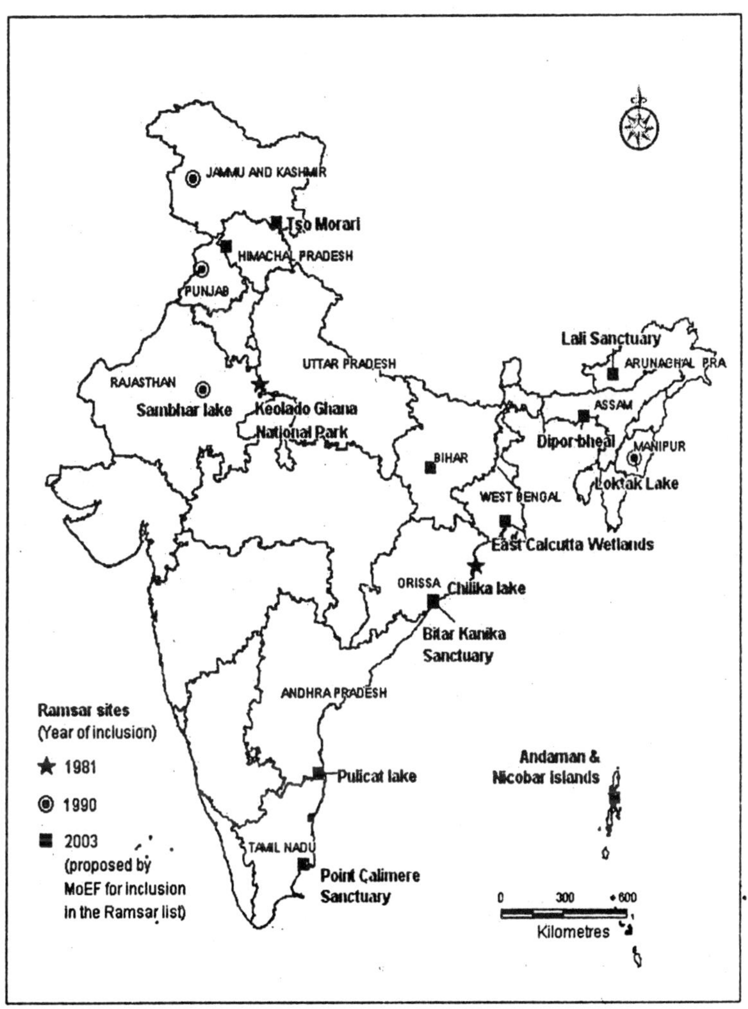

Fig. 2.5: Location of Ramsar sites in India

Wetland Eco-Restoration and Management

The proper management of any wetland requires an integrated approach that examines the dynamics of the wetland, the drainage basin and the effects of land use/land cover occurring within the watershed. The several ecological units of the wetland and its catchment that bring their own set of distinctive ecological and environmental characteristics, settlement pattern, history and the potential environmental effects, must also be considered for designing a comprehensive management strategy for the watershed.

In the subsequent sections, some of the management principles that are widely applied, both for wetlands and lakes, have been presented. Also, the role of environmental legislation and administration in protecting the wetlands has been discussed.

Wetland Legislation

There are no specific legislations meant for protecting wetlands; though there are numerous environment related legislations which provide indirect protection for them (Table 2.2). Hence the legal protection offered to wetlands need to be enhanced and the ambit of the existing environmental laws be extended to include wetland protection as well.

India is a signatory of many international conventions including CITES (Convention on International Trade in Endangered Species of wild flora and fauna, 1973), Ramsar Convention (1971), CBD (Convention on Biological Diversity, 1992), Natural Heritage Convention (1972), and others (Table 2.3). As a member of the CBD, India is committed to conserve the variety of animals and plants within its jurisdiction and to ensure that the use of biological resources is sustainable (Grimmett et al. 1998). Under the Ramsar convention, India is required to protect its wetlands that are of international importance, and foster their wise use.

Wetland Administration

The Indian Board of Wild Life (1995) and the subsequently created State Wildlife Advisory Boards (1972) act as the advisory bodies for the Central government and the State governments

Table—2.2: Some of the laws that protect Indian wetlands directly or indirectly

The Indian Fisheries act, 1857	Maritime Zone of India (Regulation and Fishing by Foreign Vessels) Act, 1980
The Indian Forest Act, 1927	Forest (Conservation) Act, 1980
Wildlife (Protection) Act, 1972	Environment (Protection) Act 1986
Water (Prevention and Control of Pollution) Act, 1974	Coastal Zone Regulation Notification, 1991
Territorial Water, Continental Shelf, Exclusive Economic Zone and other Marine Zones Act, 1976	Wildlife (Protection) Amendment Act, 1991
Water (Prevention and Control of Pollution) Cess Act, 1977	National Conservation Strategy and Policy Statement on Environment and Development, 1992

Table—2.3: Some of the international commitments of which India is a signatory

London Convention (1933) Convention Relative to the Preservation of Fauna and Flora in Their Natural State	*MARPOL* (1973/78) International Convention for the Prevention of Pollution from Ships
Ramsar Convention (1971) Convention on wetlands of International Importance Especially as Waterfowl Habitat	*Bonn Convention* (1979) The Convention on the Conservation of Migratory Species of Wild Animals.
Natural Heritage Convention (1972) Convention concerning the Protection of the World Cultural and Natural Heritage	*UNCLOS* (1982) United Nations Convention on the Law of the Sea
CITES (1973) Convention on International Trade in Endangered Species of Wild Fauna and Flora	*Convention on Biological Diversity* (CBD) (1992).

respectively in the matters related to wild life, forests, estuaries, mangroves, and general wilderness areas. The Ministry of Environment and Forests (MoEF) is concerned with panning, promotion, and coordination of all protected areas and environmental programmes. For providing sharp focus on wetland management and conservation, MoEF has created a separate 'Wetland Directorate'.

Though specific administrative machinery has been created in India for the conservation of forests and wildlife, there is no such system with a clear brief or delegation of responsibility to conserve wetlands. Only when a wetland is declared a sanctuary under the Wildlife Protection Act it is marked for conservation, albeit with typical bureaucratic tardiness and dispassion (Chari et. al., 2003). Even for such designated wetlands, there is no single multidisciplinary body charged with initiating and managing coordinated efforts to protect its different facets: water, fish, wildlife, flora, catchment outside the waterbody, etc. Given the fact that India had participated in the Ramsar convention held in 1971 and had, along with other countries pledged to conserve their wetlands and make wise use of them, this situation calls for introspection and corrective action.

Catchment Management

If the wetland or a lake is dominated by surface inflows, the characteristics of its watershed and activities occurring there play important role in determining water quality and productivity (Baker et al. 1993).

Rast and Holland (1988) mention conservation tillage, vegetative buffer strips, contour cultivation, cross-slope tillage, strip cropping, correct fertilizer application practices, tile drainage, and management of livestock manure for controlling the nutrients and sediments through run-off. Some of the options for controlling non-point source pollution include street sweeping, catchment or basin cleaning, detention basins and storm sewers.

Insitu Wetland Management

The treatment measures that can be applied directly to a wetland or lake can alleviate the symptoms of degradation rather than the cause. Also, these measures would not be effective unless proper planning, decision making and socio-economic perspectives are considered.

Planning and Decision Making: Socio-economic Perspectives

1. Wetlands are often used for multiple and even conflicting purposes. Wetland management decisions must reflect the need to set priorities, compromise, and establish an optimal balance among multiple uses.

2. Management decisions and actions must be cost effective. Cost considerations can be incorporated directly into the planning and management decisions.
3. Wetlands re complex ecosystems and involve potential adverse consequences of making mistakes. Thus, the planning and management must involve the well-trained professionals in the fields of limnology, fisheries and other related fields that involve crucial decision making.

Nutrient Control

The nutrients entering the lake can be limited by: (a) removal of nitrogen and phosphorous at their source, and (b) diverting nutrient-rich effluents or waste-waters from the receiving bodies.

Some of the 'in-situ' methods for controlling nutrients in wetlands/lakes are:

1. *Dilution.* Dilution of nutrients in lakes by the controlled addition of water that has low nutrient contents (Oglesby and Edmondson, 1966).
2. *Aeration.* Remobilization of phosphorous is higher in the anaerobic water than the well aerated waters (Baker et al., 1993). The oxygenated forms of manganese and iron form insoluble precipitate with phosphorous making it unavailable. Thus, aeration techniques increase oxygen levels in wetlands/lakes that are deep. In addition, dissolved oxygen has a positive effect on the biota encouraging the fish and beneficial organisms to grow.
3. *Oxidation of sediments.* Oxidation of the lake's sediments would help in reducing the mobility of sediment bound phosphorous into the water column. Oxidation of sediments can be done through pumping of air and by applying oxidizing agents such as calcium citrate. The latter procedure of applying oxidizing agents is termed as 'RIPLOX', after its originator Ripl (1976).
4. *Physico-chemical methods.* As high as 90% of the phosphorous and nitrogenous compounds can be

removed from the wetlands/lakes using pre-precipitation, simultaneous precipitation and post-precipitation methods (Balsrud and Balmer, 1973).

(*i*) Aluminum salts, such as aluminum sulfate (alum) and sodium aluminate, because of their strong affinity to adsorb and absorb inorganic phosphorous, remove phosphorous containing particulate matter from the water column as part of the floc (Baker et al., 1993). However, aluminum salts need to be used in appropriate doses as they are toxic for fishes and other biota. Also, zirconium oxychloride has been found to effectively precipitate phosphates (Kumar and Rai, 1978).

(*ii*) In un-nitrified, low-BOD water or effluents the chief component that consumes oxygen is ammonia. Thus, nitrification can be a very useful process for the removal of nitrogen from the effluent. Nitrogen can be removed by (a) biological nitrification and denitrification (b) air stripping of ammonia from an alkalized wastewater (c) ion-exchange (d) electrodialysis and (e) reverse-osmosis.

Nitrogen removal by air-stripping involves raising the pH of water to make it sufficiently alkaline which shifts the ammonium-ammonia equilibrium to the ammonia side, and NH_3 gas so produced is trapped in water in air-stripping towers.

5. *Sediment removal*. Nutrients, especially phosphorous, and the decaying organic particulate matter settles down in water bodies, making the sediments rich in nutrients. These sediments with high concentrations of phosphorous or nitrogen, serve as internal nutrient source. Thus, dredging and the removal of these sediments can substantially reduce nutrient recycling and availability (Olem and Flock, 1990).

6. *Biological Methods*

   (i) Reduce the amounts of nutrients solublilized in water through microbial decomposition of bottom sediments; this can often be achieved by the bottom-sealing technique of Sylvester and Seabloom (1965), i.e., artificially planting an inert layer which covers bottom sediments.

   (ii) The ammonium nitrogen can be biologically oxidized to nitrate which is in turn biologically reduced to nitrogen gas.

   (iii) It is essential first to reduce the organic content of the water by means of carbon oxidizing organisms and then to decrease its nitrogen content through the mediation of nitrifying and denitrifying bacteria.

Controlling Algal Blooms

1. Mechanical removal of higher plants and algal blooms can reduce the amount of nutrients recycled into the water upon their death.
2. The growth and multiplication of algae can be chemically controlled using copper sulfate and sodium arsenite (Kumar, 1983).
3. In recent times, the use of natural food webs, such as releasing planktivorous fish, daphnids and fishes, which can remove algae, has been widely tested with encouraging results.

Control of Aquatic Macrophytes

Excess of aquatic macrophytes are undesirable for fishing, aesthetics, boating, or swimming. The extent and density of aquatic macrophytes can be controlled using mechanical, chemical, and biological methods.

1. *Water level draw down*: Draw down of the lake water would help in controlling certain aquatic macrophytes due to desiccation. However, water draw-downs do stimulate the growth of some plants such as hydrilla

and alligator weed. In addition, water draw-downs also may be used for controlling over crowded, stunted fish populations. When water levels are lowered, the small fish populations that seek shelter in littoral weed beds are forced to enter the open water making them susceptible to predation.

2. *Shading and sediment covers:* Covers, such as polyethylene polypropylene, fiberglass or other similar materials, when placed on the water or sediment surface acts as a physical barrier to plant growth by blocking light (Engel, 1984). Even planting evergreen shady trees along the shores would cut sunlight sufficiently to reduce the photosynthetic rate of primary producers in the lake.

3. *Using insects:* The exotic insect species that infest on selective plant species is known to work best in conjunction with mechanical harvesting and use of herbicides (Olem and Flock, 1990).

4. *Using Fishes:* Of the several methods (Table 2.4) grass carp (Ctenopharyngdon idella, white amur) had been the most effective at controlling new aquatic plant growth as per Cooke et. al. (1986). The grass carp is a voracious feeder of aquatic plants particularly pond weeds (Potamogeton spp.), naids (Najas Spp.) and hydrilla.

The triploid grass carp, genetically derived from the diploid grass carp, is functionally sterile and hence minimizing the potential for uncontrolled spread of the species.

Apart from grass carp, Tilapia zilli and Tilapia aurea also feed on macrophytes and filamentous algae (Cooke et al, 1986). These species can also be considered for the possible introduction in the lake for controlling the macrophytes and algae.

Table—2.4: Comparison of lake restoration and management techniques for control of nuisance aquatic weeds (source: Olem and Flock, 1990)

Treatment (one application)	Short-term effectiveness	Long-term effectiveness	Cost	Chance of negative effects
Sediment removal	E	E	P	F
Drawdown of water	G	F	E	F
Sediment covers	E	F	P	L
Grass Carp	P	E	E	F
Insects	P	G	E	L
Harvesting	E	F	F	F
Herbicides	E	P	F	H

E = Excellent; F= Fair; G= Good; P= Poor; H= High; and L= Low

5. *Use of Herbicides.* Herbicide treatments drastically reduce macrophyte growths, and the benefits are short term with potential negative side effects. Diquat, Endothll, 2,4-D, Glyphosate, and Fluridone are commonly used herbicides.

A comparison of the various management techniques complied by Olem and Flock (1990) has been presented as Table 2.4.

Reducing Water Turbidity

High turbidity in water bodies may make them aesthetically unpleasing and reduce its desirablity for swimming. Some fish species such as common carp are known to disturb the lake bottom as they feed, increasing the turbidity.

The abundance of planktivorous fish may influence the abundance of zooplankton, which influences in turn the abundance of phytoplankton and water clarity. Also, reductions in the numbers of planktivorous fish may indirectly increase water clarity. Reductions in planktivorous fish may be achieved through removal or increased fishing pressure on planktivorous species or indirectly by increasing the abundance of larger fish predators that feed on the smaller planktivorous fish.

Do Nothing Option

The option of doing nothing sometimes helps in evaluating eutrophication management options, and to decide whether or not to implement the management programmes.

Acknowledgements

SAA thanks the University Grants Commission, New Delhi, for a major research project which supported this study. KBC thanks the Council for Scientific and Industrial Research, New Delhi, for the senior research fellowship.

REFERENCES

Abbasi S.A., 1997. *Wetlands of India -ecology and Threats: The Ecology and the Exploitation of Typical South Indian Wetlands*, Vol.I, Discovery Publishing House, New Delhi.

Ambasht R.S., Srivastava A.K., and Srivastava S.K., 1990. *Perspectives in Environment: Types and Values of Wetlands*, Ashish Publishing House, New Delhi.

Baker J.P., Olem H., Creager C.S., Marcus M.D., and Parkhurst B.R., 1993. *Fish and Fisheries Management in Lakes and Reservoirs*, EPA 841-R-93-002, Terrene Institute and U.S. Environmental Protection Agency, Washington, DC.

Balsrud K., and Balmer P., 1973. *What Ameliorating Measures are Required in Addition to the Primary and Biological Treatment Applied in Waste-water Treatment Plants*, Inform. Bulletin No.20, Europ. Fed. Protec.Waters, pp. 76-79.

Chari, K. B., Abbasi, S. A., and Ganapathy S., 2003. *Ecology, Habitat and Bird Community Structure at Oussudu Take: Towards a Strategy for Conservation and Management*, Aquatic Conserv: Mar. Freshw. Ecosyst. pp. 1: 373-386.

Cooke G.D., Welch E.B., Peterson S.A., and Newroth P.R., 1986. *Lake and Reservoir Restoration*, Butterworth Publishers., Stoneham, M.A.

Cowardin L. M., Cartor V., Golet F.C., and La Roa E.T., 1979. *Classification of Wetlands end Deep Water Habitats of the United States Fish and Wildlife Service*, FWS/OBS -79/31, Washington DC.

Dugan P.J., 1990. *Wetland Conservation: A Review of Current Issues and Required Action*, IUCN Report, Gland, Switzerland.

Engel S., 1984. *Evaluating Stationery Blankets and Removable Screens for Macrophyte Control of Lakes*, J. Aquat. Plant Manage., 22, pp.:43-8.

Grimmett R., Carol I., and Tim T., 1998. *Birds of the Indian Subcontinent*, Oxford University Press, Delhi.

Kaul V., Fotedar D.N., Pandit A.K., and Trisal C.E., 1978. *A Comparative Study of Plankton Populations of Some Typical Fresh Water Bodies of Jammu and Kashmir State* In Environmental Physiology and Ecology of Plants, Sen, D.N., and Bansal, R.D., (eds.), B.S.M Pal Singh, Deharadun, pp: 249-269.

Keddy P.A., 2000. *Wetland Ecology: Principles and Conservation,* Cambridge University Press, USA.

Kumar H.D., 1983. *Modern Concepts of Ecology,* Vikas Publishing House Pvt Ltd., New Delhi.

Kumar H.D., and Rai L.C., 1978. *Zirconium-induced Precipitation of Phosphate as a Means of Controlling Eutrophication,* Aquatic Botany, pp. 4: 359-366.

Lehmann A., and Lachavanne J. B., 1999. *Changes in the Water Quality of Lake Genewa Indicated by Submerged Macrophytes,* Fresh Water Biology, pp. 42 (3): 457-466.

Llamas M. R., 1994. *Four Case Histories of Real or Pretended Conflicts Between Ground Water Exploitation and Wetland Conservation,* Australia, pp. 1: 493-497.

Mitsch W. J., and Gosselink J.G, 1993. *Wetlands,* Van Nostrand Reinhold, New York.

MoEF Report. *http://sdnp.delhi.nic.in/resources/wetlands/articles/wetlands-moef.html*

Oglesby R. T., and Edmondson W. T., 1966. *Control of Eutrophication,* Jour. Water Poll. Contrl. Fed., pp. 38:1452-1460.

Olem H., and Flock G., (eds.) 1990. *The Lake and Reservoir Restoration Guidance Manual.* EPA 440/4-90-006, U.S. Environmental Protection Agency, Washington, DC.

Richardson C. J., 1995. *Wetlands Ecology,* In Encyclopaedia of Environmental Biology, Nierenberg W.A., (ed.), Academic Press, New York., pp: 535-550.

Ripl W., 1976. *Biochemical Oxidation of Polluted Lake Sediment with Nitrate: A New Restoration Method,* Ambio, pp. 5: 132-135.

Scott D. A., 1989. *A Directory of Asian Wetlands,* IUCN, The World Conservation Union, Gland, Switzerland, and Cambridge, UK.

Singh D. F., 1996. *An Overview of Indian Wetlands, In Assessment of Water Pollution,* Mishra S.R., (ed.), Ashish Publishing House, pp: 357-368.

Sylvester R. O., and Seabloom, R.W., 1965. *Quality of Impounded Water as Influenced by Site Preparation, Dept. Of Civil Engineering, Univ of Washington, Seattle.*

Trisal C.L., 1977. *Studies on Primary Production in Some Kashmir Lakes,* PhD Thesis, Kashmir University, Kashmir.

Watershedss. *http://h2osparc.wq.ncsu.edu/info/wetlands/onsite.html*

WWF Report. http://www. wwfindia.org/programs/fresh-wet/ramasar.jsp?prm=53

SECTION—II

THE KALIVELI LAKE

3

Kaliveli Wetland

A Typical Tulwark Against Drought

Abstract

Kaliveli is one of the most prominent among wetlands that dot the semi-arid landscape of the peninsular east coast. In this chapter we present an overview of the role played by Kaliveli in water storage, groundwater recharge, fisheries, and wildlife protection. We have also identified threats of eco-degradation and the various measures necessary to conserve the wetland.

Key words: Wetland, land use/land cover, GIS, remote sensing

Introduction

With a water-spread area of ~ 70 km^2 and a catchment spanning ~756 km^2, Kaliveli is among the 20 largest wetlands of India and is the largest of the wetlands falling within Tamil Nadu (Scott 1989, Abbasi 1997) It's petal-shaped water body lies parallel to the east coast (Fig. 3.1); the southern tip is situated about 18 Km north of downtown Pondicherry and the northern tip is ~ 85 km south of downtown Chennai. A narrow channel links Kaliveli with the Bay of Bengal via Yedayanthittu estuary. Some features of Kaliveli are presented in Table 3.1 (Chari 1997).

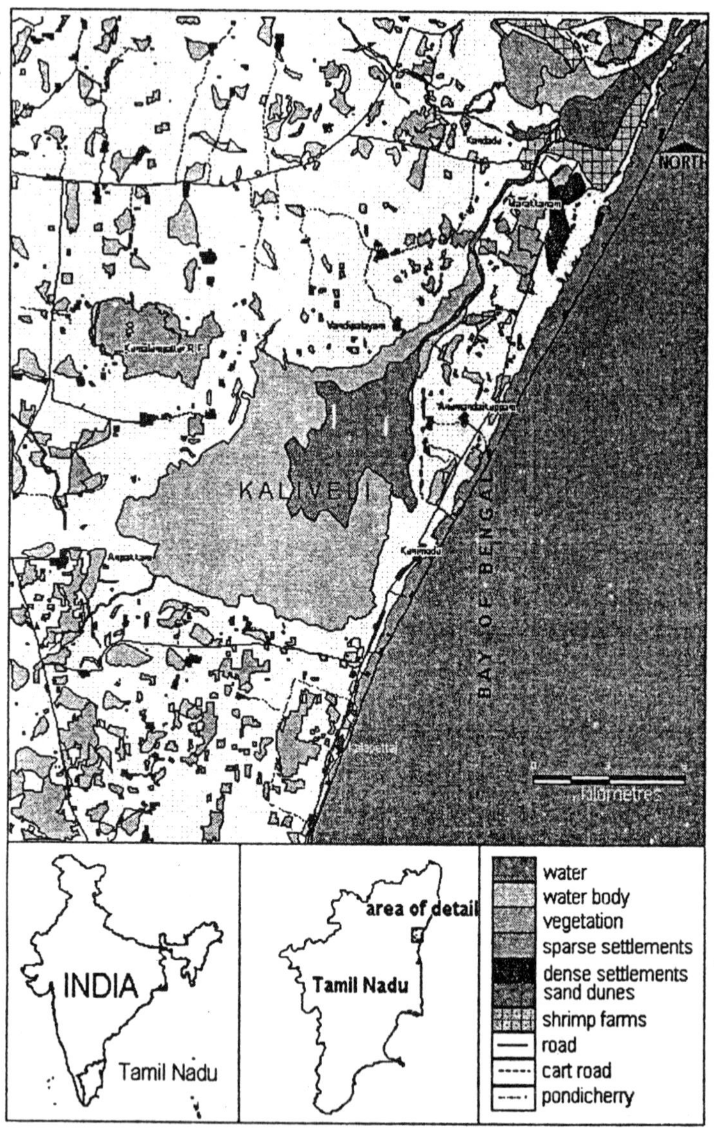

Fig. 3.1: Location and land use/land cover map of Kaliveli and its environs as per the Survey of India toposheet (1970)

Table—3.1: Some features of Kaliveli wetland

Location	Tindivanam Taluk, between Marakanam and Puthupet on the East coast, connected through a 8 km long tidal creek to Yedayanthittu estuary in the North. 12°5′-12° 10′ N and 79° 47′-79° 55′ E.
Origin	Tectonic
Rainfall	127 cm/year
Temperature	23°C-35°C
Area	70.5 Km^2
Catchment	Free catchment: 116.4 Km^2 Intercepted catchment: 636.3 Km^2 Combined catchment: 754.7 Km^2 Equivalent catchment: 244.0 Km^2
Storage capacity	34 million m^3
Average depth	3 ft
Maximum depth	6 ft

Water Spread

Kaliveli is a typical bulwark against drought in a semi-arid landscape. It helps in catching rainwater falling over ~ 755 km^2, and forms a ~ 70 km^2 span of water which recharges groundwater for most of the year. Kaliveli also supports irrigation, domestic water needs, and fisheries. And it forms a buffer against seasonal storms (Chari 1998).

Kaliveli fills up to its maximum extent by the end of the Northeast monsoon, (Fig. 3.2a). Its water-spread shrinks in subsequent months (Fig. 3.2b, c). In the years when there is no or very low rainfall during January-June, the wetland can dry out completely (Fig. 3.2d). The average depth of water at the end of the northeast monsoon is about one metre, and the maximum depth about two meters. By the end of the monsoons, the lagoon is normally full of fresh water having received copious run-off from neighbouring farmlands and other parts of its watershed. Subsequently, as the run-in of fresh water diminishes, there is some inflow of seawater from the estuary, and the lagoon becomes brackish, at its northern end.

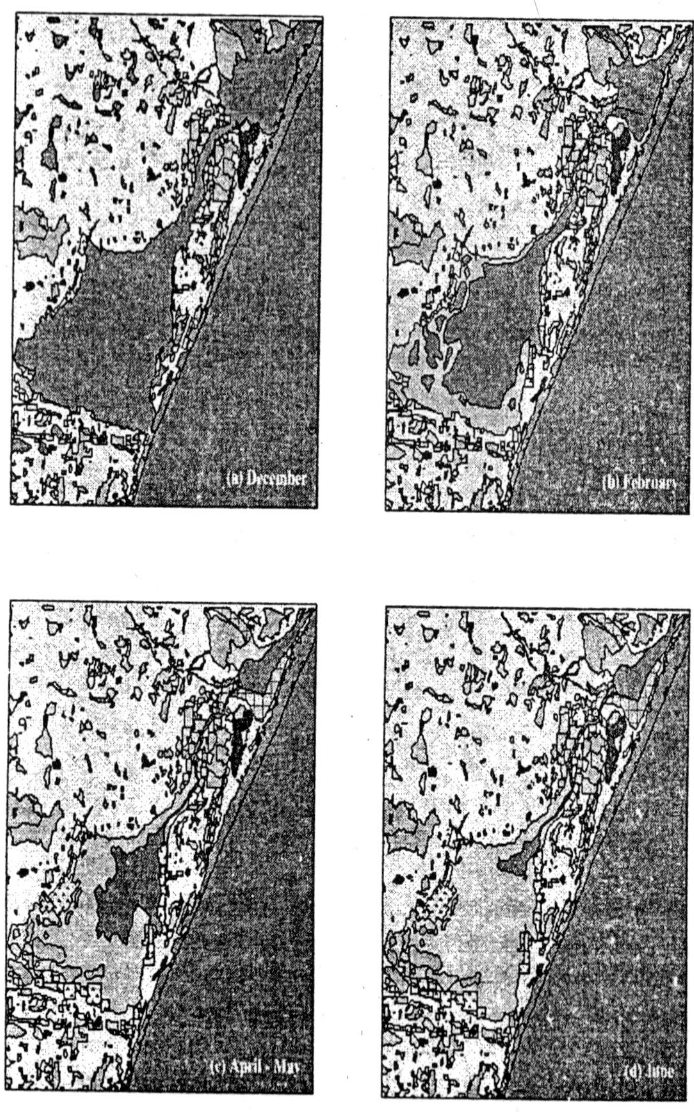

Fig. 3.2: The water spread of Kaliveli wetland along the various months

The benefits accruing from Kaliveli, the threats to the wetland's sustainability, and some pointers for further action are summarized below:

Wetland Values of Kaliveli

Kaliveli, because of its sheer vastness, is known to recharge the ground water aquifers of its bioregion (Abbasi, 1997; Chari 1998. The presence of this wetland in the proximity of the sea adds much more significance as it checks the saline water intrusion into the ground water aquifers.

The fresh water part of Kaliveli supports extensive fishing when the water is present during the months of Northeast monsoons. As the Yedayanthitu estuary has water throughout the year due to the influx of sea water, fishing is done throughout the year.

Kaliveli is highly eutrophic and thus supports extensive reed beds and grasses. As the water in the wetland starts drying up, the reeds and grasses are harvested for the purpose of thatching the huts, firewood, and fodder for domestic animals.

The brackish part of the wetland, near Vadagaram, Mudaliarpet, and the Yedayanthitu estuary is used for shrimp farming, and salt manufacturing.

Some portions of the wetland, towards Northeast support India's third largest salt marshes (Chari 1997).

Threats

The water-spread of Kaliveli is shrinking rapidly due to encroachment by paddy fields and salt pans. The intensive practice of agriculture around Kaliveli can contaminate the wetland with pesticides and nutrients. Poaching of birds by several means such as nets, poisoned fish as bait and shooting is quite common.

Some incidents have been reported of industries dumping their effluents and wastes in-and- around the wetland. The release of effluents by the caustic soda plant, located some five km down south of Kaliveli, into the sea may backlash into the wetland.

The Imperatives

Several organizations, including French Institute of Pondicherry, Pondicherry University, and NGOs such as COPDANET (Coastal Planning & Development Action Network) have recommended that Kaliveli be made into a bird sanctuary. Re-afforestation of mangrove vegetation has also been recommended. Some of the significant pointers are:

1. The wetland and the entire watershed have also been recommended for designation as a Biosphere Reserve under the UNESCO *Man and Biosphere (MAB) Programme.*
2. The development of an educational programme to demonstrate to the local people that management of Kaliveli will be in their long term interest.
3. A detailed study on the vegetation structure and dynamics of the wetland is necessary.
4. Planting of trees around Kaliveli would provide nesting habitat for water-birds.
5. There has been a controversial suggestion to develop Kaliveli as a tourist spot focusing on conservation and a bird sanctuary.
6. It is believed that Kaliveli watershed provides an ideal site for a model study of integrated watershed development.

The management plans for Kaliveli should take into account the diverse uses made by villagers of this wetland. Traditional activities such as fishing, reed cutting, grazing, de-silting etc., if periodically monitored to prevent overuse, can cause little disturbance to the wetland. Indeed ecosystems like Kaliveli get adjusted to moderate 'disturbances' that humans have been causing over centuries; an equilibrium or stabilization gets achieved.

REFERENCES

Abbasi S.A, (1997), Wetlands of India-ecology and Threats: the Ecology and the Exploitation of Typical South Indian Wetlands, Vol.I, Discovery Publishing House, New Delhi,p. 149.

Chari K.B, (1997), A Reconnaissance of the Ecology of Two Wetlands: Ousteri and Kaliveli, MS., Thesis, Pondicherry University, p. 110.

Chari K.B, (1998), Environmental Status of a Shallow Fresh Water Lake South Indian Peninsula, M.phil Thesis, Pondicherry University, Pondicherry. India, p. 113.

Scott D. A, (1988), A Directory of Asian Wetlands, IUCN, Gland & Cambridge, p. 487.

SECTION—III

LAND USE PATTERN AND SOIL QUALITY OF KALIVELI AND ITS SURROUNDINGS

4

Assessing Land Use/Land Cover of Kaliveli Watershed Using Remote Sensing and GIS

*Implications for Conservation**

Abstract

Several features of the Kaliveli wetland were studied by synthesizing the inputs from satellite imagery, topo-sheets, and extensive on-ground surveys. A geographical information system (GIS) was developed for the wetland with the aid of computer-automated tool MapInfo Professional. With the aid of GIS the implications of the present land use on the lake and its water system have been identified. Various measures for conservation of the lake and its watershed have been suggested based on the above findings.

Introduction

Landuse in the watershed directly affects both quality and quantity of the water in a lake. If the lake is dominated by surface

---

* *Reproducd from the proceedings of 'The international conference on remote sensing and GIS/GPS', Hyd, Feb' 2002.*

inflows, the characteristics of its watershed and activities occurring there play important roles in determining water quality and productivity (Baker et al., 1993).

In the recent times, both remote sensing and GIS have evolved as important tools in the management of watershed, wetlands and lakes. Several authors have made efforts towards developing management systems using remote sensing and GIS such as: Baban (1999), Lehmann and Lachavanne (1999), Line et. al. (1997), Yang et. al. (1999) and Richards and Kump (1997). In this chapter we present a study on the plankton of Oussudu lake exploring the basic concepts of GIS in establishing the relationship with land use characteristics.

Methodology

Information from the Survey of India topo-sheet (No 59 P/ 16), satellite imagery and extensive ground-truth survey were synthesized to generate the maps used in the present analysis.

The satellite image, IRS-IC LISS III, dated 17 Feb 1997 was traced from the archives of Institute of Remote Sensing, Anna University, Madras. The traced image was scanned, and digitized using desktop GIS software-MapInfo professional (version 5.5). The image was later classified for the land cover/land use categories as per the system adopted from Eugene et. al., (1992).The image was interpreted by means of visual observation and substantiated by post field training/test areas.

For the purpose of ground-truth survey, all the basic features such as major water-bodies, settlements, and road network were extracted and enlarged from 1: 50,000 to 1: 25,000 scale from the topo-sheet. Ground-truth surveys were conducted on foot by a three-member team over a period of three years. The detailed parcels of information recorded during this direct survey were then integrated with the features of the topo-sheet.

Results and Discussion

The land cover map of the study area, classified as per the color codes recommended by International Geographical Union, is presented in Fig. 4.1. The satellite image has registered the

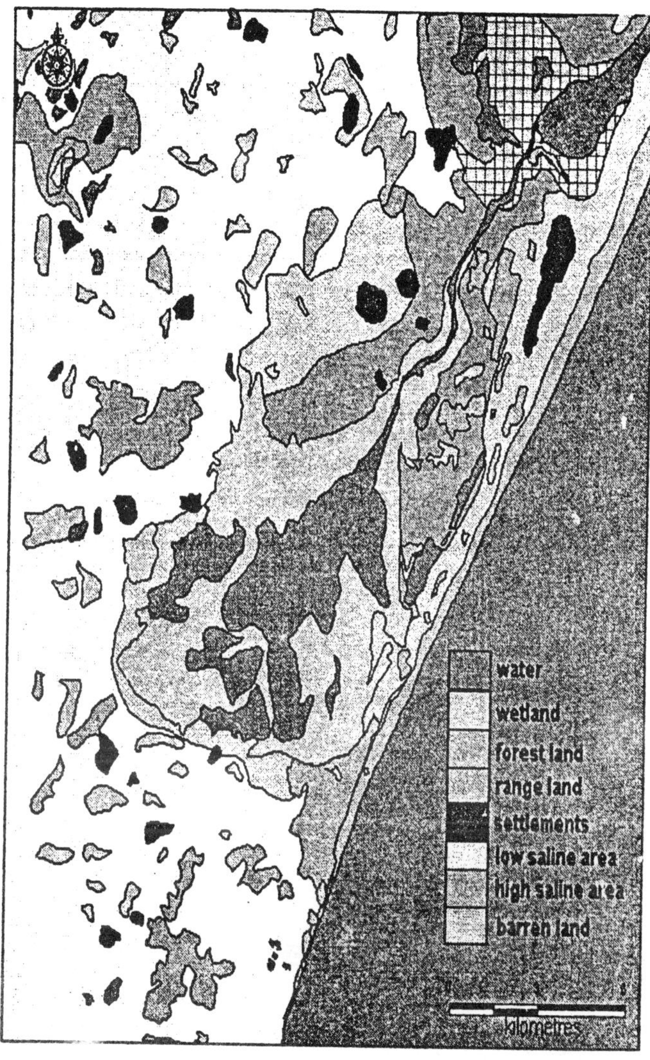

Fig. 4.1: Classified IRS-1C Liss III satellite image of Kaliveli watershed

reflectance of various land cover categories of the region at 2 (green), 3 (red) and 4 (near infrared) of electromagnetic spectrum. The region features diverse land use/land cover categories, prominent being water resources, vegetation, settlements, and barren lands (Table 4.1).

On the east coastal-marine land forms with levelled coastal plain are followed by sand dunes and beach sand with small-localized marshes along the sides.

Along the other direction of Kaliveli there is a rolling undulating kind of land form with a few localized mounds of about 100 ft height and impediments with middle and lower slopes before the land slopes into the lake. The lake as such is surrounded by a typical lacustrine landform. All over the terrain of the lacustrine plain small tanks with a network of drainage channels are found. The drainage pattern is dendritic (Scott 1989).

Table—4.1: Land cover during 1970 (as per the topo-sheet) and 1997 (satellite image)

	Land cover category	*1970*		*1997*	
Level 1	*Level2, level 3, level 4*	*Area (km²)*	*%*	*Area (km²)*	*%*
Water bodies		335.01	43.6	350.00	45.5
	Sea	219.60	28.6	219.60	28.6
	Wetlands	115.41	15.0	130.40	17.0
	Salt pans	5.49	0.7	14.90	1.9
Vegetation		–	–	–	–
	Scrubs	46.16	6.0	29.16	3.8
	Dense scrubs (forests)	11.51	1.5	8.95	1.2
	Open scrubs (range-lands)	34.65	4.5	20.21	2.6
Built-up-structures	Settlements	12.46	1.6	10.60	1.3
	Dense settlements	5.90	0.8	–	–
	Sparse settlements	6.56	0.8	–	–
	Transportation (km)	207.33	–	–	–
	Road	35.29	–	–	–
	Cart road	72.04	–	–	–
Others		375.17	48.8	379.04	49.3
	Barren lands (sand dunes)	15.49	2.0	23.03	3.0
	Others (unidentified)	359.68	46.8	356.01	46.3
Total		768.80	100	768.80	100

The lake is connected to the sea by Yedayanthithu estuary, which is surrounded by large areas of intertidal mud flats. There are some 500 hectares of salt pans along the estuary immediately across the channel of Kaliveli and near Erapanni, just one-and-half kilometer from Marakkanam bus stand. The levelled coastal plains support extensive agriculture.

Natural features: Natural features include water, watershed, vegetation, and barren land. Among the natural features water-bodies dominate with 350 km² of the total 768.80 km² stretch. Kaliveli lake and other water bodies, seen scattered throughout the study area, occupy about 115 km² (15%) of the total landcover (Fig. 4.2).

Both range lands and forest cover represent approximately 29.16 km². It was observed that the Marakanam reserve forest was almost lost, while Kumalumpattu and Agaram reserve forests have been reduced by as much as 18.7% and 28% respectively, over the years (1970-97).

The rangeland category can be located around Kaliveli. Most of it is interpreted as plantations. The plantations towards southeast of Kaliveli were found to be more dense considering the tonal quality. The field visits revealed that the plantations were chiefly that of Eucalyptus sp., Casuarina sp., and *Anacardium sp.* (cashew). Certain parcels of information such as sparse vegetation cover seem to have not been preserved by the image.

Beaches, stony waste, uncultivated lands and salt pans which fall under the classified barren land category occupy about 3% of the landcover.

Cultural features: These include settlements—both urban and rural, 'kuchcha' pathways, roads, agriculture lands and salt pans. A few congregated settlements found in the study area include Marakanam, Koonimedu, Thenpakkam, Mannur, and Tailapuram. The settlements represent an area of 8.153 km² (% of total landuse).

Vast stretches of the estuary (about 14.90 km²) at Erapanni, Marakanam, and Thenpakkam have been converted to salt pans.

Special features: Around Kaliveli very high reflectance area of 98.38 km² was identified as saline soils. The formation of saline soils could be attributed to (i) efforescence, a phenomenon that

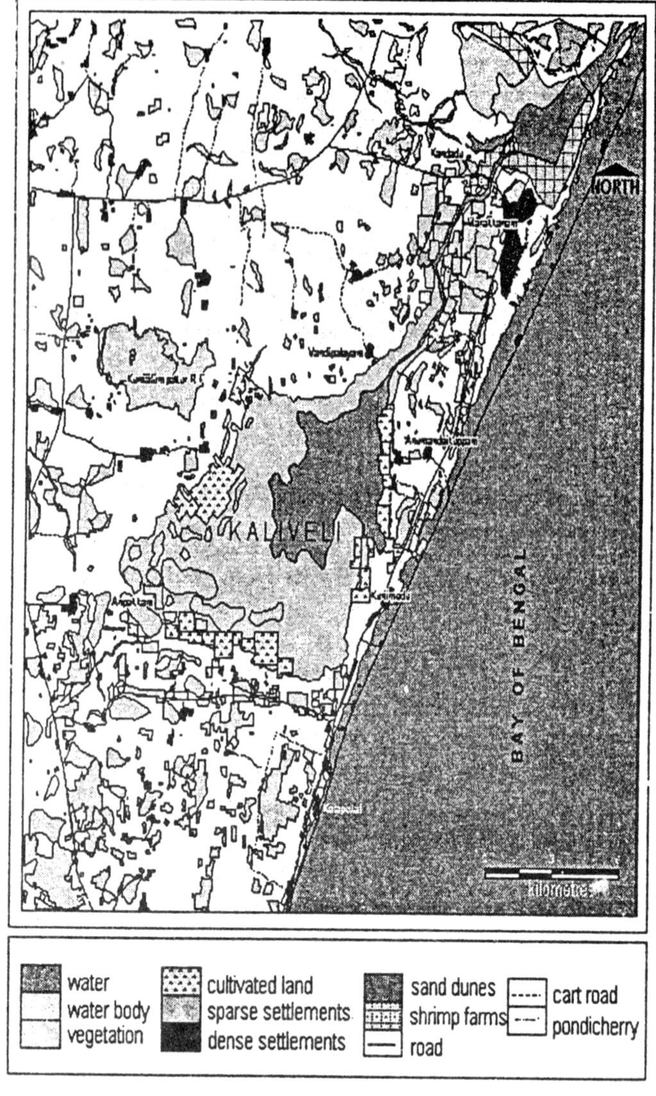

Fig. 4.2: Details of ground truth survey (conducted during 1999) integrated with the Survey of India topo sheet.

occurs in arid climates with poorly drained areas or basins having interior drainage (Garg 1980, Briggs et. al. 1997) (ii) addition of substantial amounts of soil amendments for the practice of agriculture and (iii) deposition of salt on the land, both by precipitation and dry fall out, due to the proximity of the sea (Todd 1997).

Ground-truth Survey

The ground-truth has added the much-needed information at micro-level such as cultivated lands, the crop types and other speculative land use categories that were omitted in the Survey of India toposheet. A representative study has been presented below:

The levelled plains seem to support extensive agriculture. As mentioned earlier in the chapter 3, rice paddies have encroached nearly 6 km^2 of Kaliveli (Fig. 4.3 and 4.4). Also, some portions of the lake near Aruvadi (Plate 4.1), have been leased by the government to the farmers for agriculture on a paltry rental of Rs.75 per annum! The fertilizers and pesticides from these rice paddies can pollute the lake through run-off.

Shrimp farms were sited near Agaram and Kandadu (Fig. 4.5), just beside the Yedayanthitu estuary. There is a high risk of pollution due to shrimp farms as the effluents, rich in BOD, COD, and other agents of eutrophication, find their way into the lake.

During the early monsoons the lake gets flooded and no agriculture is practised around the lake. After the water level recedes, the lands are flushed well with bore well water and left for drying. A complex (composite) of Urea and Cowdung (help decreasing the pH of soil) is applied for making the soil irrigable (Chari 1997). In some villages like Anumandaikuppam the fields were raised to half-a-meter and then agriculture is practised.

The major crops and their varieties grown around Kaliveli are:

(a) Rice Variety: IR36, IR50, IR20, IR38, IR37, IR43, white Ponni, red ponni, emergence NLR (Salinity tolerant and pest resistant plant).

(b) Groundnut Variety: JLR, NLR, IR-Red.

Plate 4.1: Part of the Kaliveli water-body, towards southwest, has been illegally cultivated and some part has been leased by the government to farmers (the portion of the land lying in front of the man standing in the picture) at the unbelievably low sum of Rs. 75 per annum! The combined impact of these actions is witnessed, as above, by the conversion of the water-spread area into dry land.

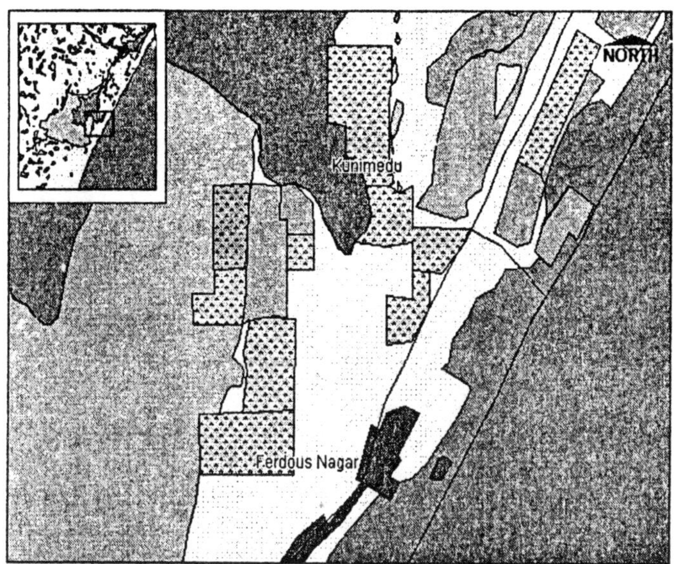

Fig. 4.3: Map showing the ground-truth detail of Kunimedu and Ferdous Nagar village (note the encroachments by agricultural fields)

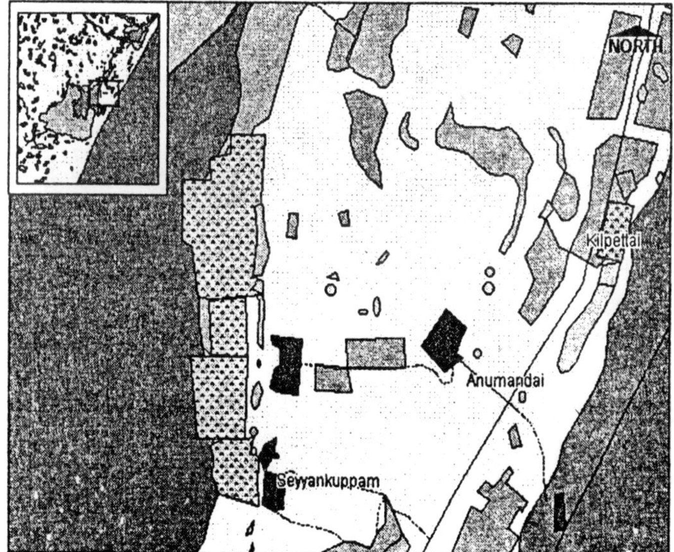

Fig. 4.4: Map showing the ground-truth detail of Anumandai village (note the agricultural fields in the precints of Kaliveli)

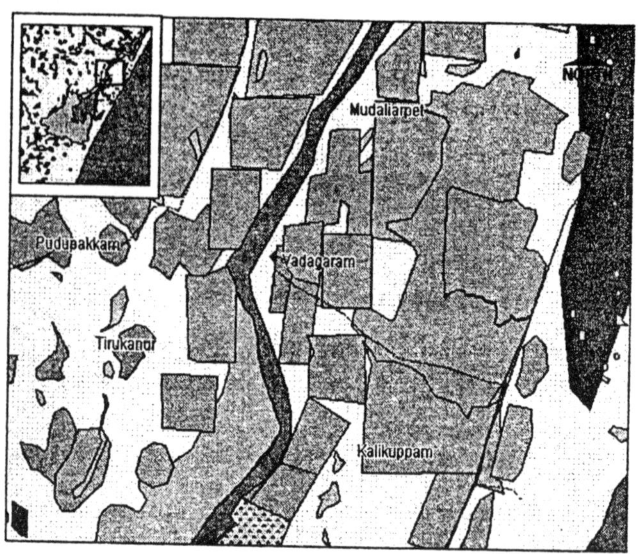

Fig.4.5: Map showing the ground truth detail of Kalikuppam, Vadagaram, and Tirukanur villages. The ground-truth detail of Kandadu and Erapanni (Note the spread of saltpans and shrimp farms).

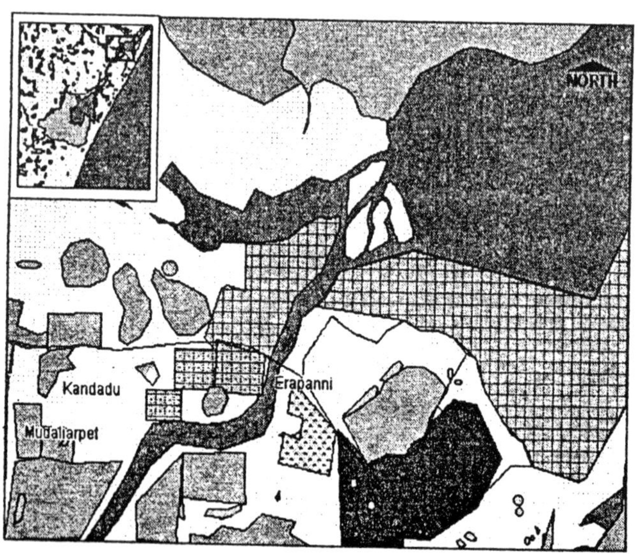

Fig.4.5(b)

(*c*) Ragi Variety: Grown after the water recedes in Karikuppam Saarada -60, Makkal,

(*d*) Cumbu Varieties: A profitable crop to grow in summer, also require no watering. Cambi-47, and White Cumbu.

(*e*) Sugarcane: It is not grown due to fox menace

(*f*) Coconut: More profitable and is used as mixed cropping.

(*g*) Bamboo: Require no maintenance

(*h*) Casuarina: More profitable and require less maintenance.

(*i*) Cowpea: Grown as intercrop with groundnut

(*j*) Bengal gram: Grown as intercrop with groundnut.

Kunimedu

Kunimedu lies at about 5 kms from the northern end of the lake. The main land use of this area is Agriculture followed by settlements. Settlements are away from the lake. Major settlements are seen at Ferdous Nagar with a density of about 200 houses.

Nearly 548.30.5 hectares of the land at Kunimedu is under cultivation. Out of which dry land crop covers 258.03.5 hectares and wetland crops cover 290.27.0 hectares.

The different types of crops grown and their coverage are given in the Table 4.2.

Encroachment of the lake by agriculture was observed to some extent.

Table—4.2: Land in Koonimedu under different agricultural practices

Dryland		*Wetland*	
Crop	*Area (ha)*	*Crop*	*Area (ha)*
Rice	73.02.3	Rice	212.15.0
Groundnut	85.01.0	Bengal gram	43.06.0
Bhendi	9.00.1	Green gram	27.04.0

(Contd...)

Coconut	65.00.1	Cotton	8.02.0
Cumbu	15.00.0		
Mustard	7.00.0		
Chilly	4.00.0		
Total	258.03.5		290.27.0

Anumandai

Anumandai consists of four hamlets: Anumandaikuppam, Kaliyankuppam, Seyyankuppam and Settikuppam.

The land coverage shows that the major land use around this area is agriculture followed by settlements.

A continuous stretch of settlements is seen from East Coast Road towards Kaliveli lake. Agriculture covers around 533.24.5 hectares of land, out of which 370.11.0 hectares of lands are under dry land agriculture and 163.13.5 hectare of lands are under wetland agriculture. Different types of crops grown and their coverage in dry land and wetland are given in Table 4.3.

Table—4.3: Land in Anumandai under different agricultural practices

Dry land		*Wet land*	
Crop	*Area (ha)*	*Crop*	*Area (ha)*
Rice	63.03.0	Rice	133.07.3
Groundnut	165.05.0	Greengram	12.00.1
Coconut	95.02.0	Bengalgram	18.06.1
Ragi	35.01.0		
Total	370.11.0		163.13.5

Karikuppam

Agriculture, settlements and sand quarrying are the major landuse types in this village. Large stretches of land besides the lake are under agriculture. However, no encroachment of the lake by Agriculture field is seen in this village.

In this village agricultural lands accounts for around 190.03.2 hectares of land. Out of which 105.03.0 hectares of lands are under dry land agriculture and 85.00.2 hectares of land are under wetland agriculture. The major crops grown and their coverage under dry and wetland agriculture are given in Table 4.4.

The Table 4.4 shows that rice is the only wetland crop grown in this village. A majority of the agricultural land is covered by rice and groundnut.

Table—4.4: Land at Karikuppam under different agricultural practices

Dry land		*Wet land*	
Crop	*Area (Ha)*	*Crop*	*Area (Ha)*
Groundnut	68.02.0	Rice	85.00.2
Coconut	14.01.0		
Ragi	23.00.0		
Total	105.03.0		85.00.2

Few settlements are seen near the lake unlike Anumandaikuppam. Settlements increase as one move away from the lake towards East Coast Road.

A sand quarry is seen near the lake. The sand extracted from this place is used for manufacturing ceramics and bottles for beverages.

Kandadu (W)

Kandadu is located to the north of Kaliveli, near Yedayanthitu estuary. The major landuse types of this area are agricultural and settlements.

The land coverage under agriculture in this village is 588.27.0 hectares of this dry land accounts for 529.16.0 hectares of land and wet land accounts for 59.11.0 hectares of land. The major crop grown in this area are groundnut, coconut and rice.

The different crops grown under dry land and wetland and its coverage is given in Table 4.5. Groundnut, coconut and Rice accounts for 3/4th of the land under agriculture.

Table—4.5: Land at Kandadu: (W) under different agricultural practices

Dry land		*Wet land*	
Crop	*Area (Ha)*	*Crop*	*Area (Ha)*
Groundnut	345.07.0	Rice	59.11.0
Coconut	108.03.0		
Bengalgram	42.02.7		
Chilly	9.02.3		
Bhendi	3.01.0		
Sugarcane	22.00.0		
Total	529.15.10		59.11.0

Around 1,000 households are there in this area and around 7000 population. Settlements are seen around the lake in clusters and are half-a-kilometer away from the lake.

Erapannai

Erapanni is also known as Kandadu (East). Agriculture, shrimp farms and saltpans are the major land use types of this area.

Agriculture accounts for around 466.28.5 hectares of land, out of which 343.11.0 hectares of land are under dry land agriculture and 123.17.5 hectares of land is under wet land agriculture. The crops grown under dry land, wetland and their coverage are given in Table 4.6.

Table—4.6: Land at Erapannai under different agricultural practices

Dry land		*Wet land*	
Crop	*Area (Ha)*	*Crop*	*Area (Ha)*
Groundnut	263.03.0	Rice	117.12.5
Coconut	52.06.0	Bengalgram	6.5.0
Cumbu	16.01.0		
Sugarcane	12.01.0		
Total	342.11.0		123.17.5

This shows that the major crop grown at Erapanni is groundnut and rice. Around five shrimp farms are located in this area.

Saltpans cover a vast area on the side of bridge toward Marakkanam. Since the soils and water are saline, this area is used for salt extraction from water.

Thenpakkam

Thenpakkam is nearer to Erapannai. Since the land and water is saline, the most prominent land use in this area is saltpans. This area also anchors a small patch of mangroves.

Conclusions

The following conclusions can be drawn from the above study:

(i) If we had assessed the situation purely on the basis of the imagery, little difference would have been discerned between 1970 and 1997. But the ground-truth studies reveal that the situation has changed drastically. The items of information not revealed by the imageries but confirmed by the ground survey are:

  (a) encroachment and practice of agriculture in and around Kaliveli

  (b) mushrooming of shrimp farms near Erapanni and Mudaliarpet.

These trends if not checked immediately can prove detrimental to the lake on the long run.

(ii) High and low saline areas occur in the vicinity of the lake due to waterlogged conditions, practice of intensive agriculture, and dry/wet deposition of salt because of the proximity of the sea.

(iii) The lake and its potential water spread area need to be earmarked to check further encroachment.

(iv) Considering the ecological significance, as substantiated by the present and the earlier studies, Kaliveli should be declared a biosphere reserve.

REFERENCES

Abbasi S.A. (1997). *Wetlands of India-Ecology and Threats: The Ecology and the Exploitation of Typical South Indian Wetlands*, Vol.I, Discovery Publications House, New Delhi, p:149.

APHA (1995). *Standard Methods for Examination of Water and Waste Water*, 19th edition, American Public Health Association and American Association and Water Pollution Control Federation , Washington, U.S.A.

Baban S. M. J, (1999). *Use of Remote Sensing and Geographic Information Systems in Developing Lake Management Strategies*, Hydrobiologia, Vol 396, pp. 211-226.

Baker J.P, Olem H, Creager C.S, Marcus M.D, and Parkhurst B.R, (1993). *Fish and Fisheries Management in Lakes and Reservoirs*. EPA 841-R-93-002. Terrene Institute and U.S. Environmental Protection Agency, Washington DC.

Briggs D., Smithson P., Addison K., and Atkinson. K. (1997). *Fundamentals of the Physical Environment*, Routledge, London and New York, pp : 335 and 336

Chari K. B. (1997). A Reconnaissance of the Ecology of Two Wetlands: Ousteri and Kaliveli, MS. Thesis, Pondicherry University, Pondicherry, India, p:108

Eugene Avery T., Berline G. L., (1992). Fundamentals of Remote Sensing and Air-photo Interpretation, Fifth Edition, Macmillan Publishing Company, New York, pp: 201 -224.

Garg S. K. (1980). *Irrigation Engineering and Hydraulic Structures*, Khanna Publishers, Delhi, p :1150.

Lehmann A and Lachavanne J.B, (1999). *Changes in the Water Quality of Lake Geneva Indicated by Submerged Macrophytes*, Fresh Water Biology, Vol : 42, Iss 3, pp.457-466.

Line D.E, Coffey S.W and Osmaond D.L, (1997). *WATERSHEDS GRASS-AGPNS Model Tool*, Transactions of the ASAE, Vol 40, Iss 4, pp. 971-975.

Richards P.L, and Kump L.R, (1997). *Application of the Geographical Information Systems Approach to Watershed Mass Balance Studies*, Hydrological Processes, Vol 11, Iss 7, pp.671-694.

Scott D.A., (1989). *A Directory of Asian Wetlands*, IUCN, Gland & Cambridge, p. 487.

Todd D. K., (1997). *Ground Water Hydrology*, John Wiley and Sons, Inc., Singapore. p: 535

Yang M.D, Merry C.J and Sykes R.M, (1999). *Integration of Water Quality Modelling, Remote Sensing and GIS*, American Water Resources Association, Vol 35, Iss 2, pp. 253-263.

5

Soil Quality of Kaliveli and Its Surroundings

Abstract

This chapter deals with the soil quality of Kaliveli region, studied vis-a-vis the soil texture, pH, electrical conductivity and the nutrient status. Clay was the predominant soil type found in the study area. As clay is known to bind positively charged cations, the contribution of the catchment in terms of the soil cations-Ca^{2+}, Mg^{2+}, K^+, Na^+, and H^+, is likely to be negligible. On the other hand, the negatively charged cations, such as Cl^{-1} and No^{3-} , which are not bound by this soil, are likely to be abundant in the run-off.

A majority of the soil samples of Kaliveli region were of neutral *pH* regime. The acidic soil samples identified in the study area could be attributed to the soil acidification influenced by the continual use of inorganic fertilisers on agricultural soils.

The saline soils identified at Kaliveli can be related to: (i) the over extended period of evaporation and evapotranspiration of soil exceeding the downward percolation of rainfall, irrigation water, (ii) 'salt spray' due to the proximity of sea and (iii) 'efflorescence'

In general, the phosphorous levels appear to be negligible when compared to the nitrogen. The relatively high total nitrogen content in some of the soils could be due to the extensive agriculture practice and mineralization of organic substances in those areas.

Introduction

Soil, alongside water and biota, forms one of the three most important and dominant dimensions of any ecosystem. Besides supporting most of the plant and animal life, soil also acts as a natural 'filter' for rainwater percolating down the earth. In this chapter soil quality of Kaliveli and its catchment has been presented in terms of soil texture, pH, electrical conductivity and nutrient status.

Materials and Methods

Sampling

Soil samples were collected from 22 sites located in-and-around Kaliveli during the peak summer of June 2001 (see fig. 5.1). A quadrat of 15cm x 15cm x 15cm was placed at random in the field. After digging, the soil contents were collected in plastic bags and labelled appropriately. All samples were air-dried by placing them in a shallow tray in a well ventilated place until the soil contents have become dusty. Then the soil lumps were crushed, powdered and sieved through 2mm sieve for analyses of physical characteristics, nitrogen and phosphorous. The soil samples were further sieved through a 60 mm.

Analytical Methods

The analytical procedures adopted are as detailed in the laboratory methods of soil and plant analysis by Okalebo et.al. (1993) and APHA (1998).

Soil Texture: Soil texture is a collective term that defines a real soil by the proportion of different particle size components. For identifying the soil texture, a known amount of soil was picked by coning and rejection method and placed in a measuring cylinder with addition of water. The contents were stirred thoroughly and left undisturbed for 24 hours. The percentage of various soil particles, sand, silt and clay were measured converting the 'ml' of the measuring cylinder into percentage. Once the sand, silt and

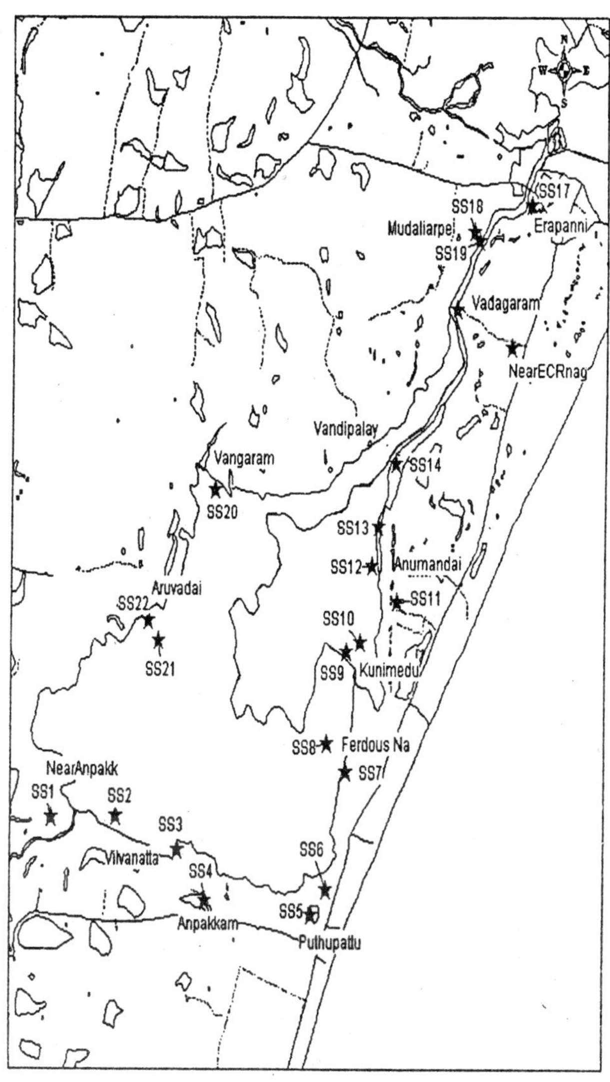

Fig. 5.1: Location of soil sampling stations in-and-around Kaliveli

clay distribution was increased, the soil was assigned to the texture class based on the soil textural triangle, Dispersion of the soil particles into the individual particle and the settling velocity estimates the percentage of sand silt and clay.

pH and Electrical Conductivity: A 2.5:1 ratio of soil and water was taken out and the mixture was stirred for 10 minutes. After allowing the contents to stand for 30 minutes the pH of the soil suspension was estimated potentiometrically using the pH meter. After allowing the contents to settle for 1 hour the conductivity of the supernatant was measured. For samples with an EC greater than 1.0 ms/cm saturated paste extract method (Okalebo et al, 1993) was followed.

Soil Nutrients: Analysis of total nutrients was done by means of complete oxidation of organic matter. This is accomplished through dry ashing or acid/alkaline digestion of the material in question. Wet acid oxidation is based on a Kjeldahl oxidation to leave a sulphuric acid solution. Hydrogen peroxide is added as an additional oxidising agent, selenium is used as a catalyst while lithium sulphate is added to raise the boiling point of the mixture. The main advantages of this method are that single digestion is required (for either soil or plant material) to bring nearly all nutrients into solution; no volatilisation of metals, N and P takes place and the method is simple and rapid (Okalebo, et. al. 1993).

Total Nitrogen: Acid digestion of the soil material was followed by Kjeldahl distillation. The distillate was collected in boric acid solution. The total nitrogen was estimated by colorimetry using Nessler's reagent as per standard methods (APHA, 1998).

Total Phosphorous: The sample digest solution was diluted appropriately and the phosphorous content was estimated by colorimetry (Sn_2Cl method) as per the standard methods (APHA, 1998).

Results and Discussion

Soil Texture

Nearly 40% of the soil samples collected (Table 5.1) in-and-around Kalveli (Figure 5.1) fall under the category of clay. Clay, according to Brusseau and Bohn (1996), are predominantly negatively charged soil components. Cation exchange is a major sorption mechanism for inorganic pollutants. Generally positively charged ions are attracted to the negatively charged sites on the clay particles. Thus, the common soil cations such as Ca^{2+}, Mg^{2+}, K^{+}, Na^{+}, and H^{+} can exist predominantly on the exchange sites or in the soil solution. The anionic ions being negatively charged are not attracted by the soil particularly thus making them mobile (Cl^{-1}, No^{3-}).

Table—5.1: Soil quality in the vicinity of Kaliveli wetland, June 2001

Sam No.	*Location*	*Soil texture*	*pH*	*EC (mmhos)*	*Total phosphorous (%)*	*Total nitrogen (%)*
SS1	NearAnpakkam	Sand	6.2	0.68	0.0073	0.019
SS2	Vilvanattam	Silty loam	4.3	0.92	0.0025	0.025
SS3	Koluvari	Sandy clay loam	5.3	22.74	0.0034	0.014
SS4	Anpakkam	Clay	6.1	13.63	0.0070	0.075
SS5	Puthupattu	Silty clay	6.8	17.77	0.0051	0.107
SS6	Puthupattu	Clay	7.2	13.75	0.0099	0.017
SS7	Ferdous Nagar	Clay	6.1	14.41	0.0013	0.024
SS8	Ferdous Nagar	Silty clay	5.1	3.00	0.0056	0.100
SS9	Kunimedu	Silty clay	7.4	23.26	0.0031	0.109
SS10	Kunimedu	Loamy sand	4.8	0.75	0.0047	0.073
SS11	Seyyankuppam	Loam	7.1	22.68	0.0050	0.106
SS12	Anumandai kuppam	Loam	7.1	5.47	0.0072	0.035
SS13	Anumandai kuppam	Sand	6.3	0.30	-	0.036
SS14	Vandipalay	Silty clay	6.5	18.97	-	0.072
SS15	Near ECR Nagar	Silty loam	6.3	0.24	0.0091	0.048
SS16	Vadagaram	Clay	4.7	0.95	-	0.059
SS17	Erapanni	Clay	7.9	15.26	0.0041	0.050
SS18	Mudaliarpet	Sandy loam	7.2	0.48	-	0.080
SS19	Mudaliarpet	Clay	6.9	3.13	0.0030	0.088
SS20	Vangaram	Clay	7.2	0.93	0.0032	0.066
SS21	Aruvadai	Sandy clay	5.0	0.23	0.0091	0.122
SS22	Aruvadai	Sandy clay	7.4	20.69	0.0080	0.120

Sand and sandy loams were identified at Anumandaikuppam and Seyyankuppam where sand quarrying is a common practice. While Silt and silt loams were identified at site SS2 and at site SS3.

Soil pH

Nearly 72% of the soil samples fall under the neutral pH regime (Fig. 5.2a) and 27% acidic (Table 5.2). The acidic soils identified at Icchangadu and Koluvari (SS3) could be attributed to the soil acidification influenced by the continual use of inorganic fertilisers on agricultural soils. The adverse effects of such soils on plant growth are two folds. Extreme acidity can give rise to high solubility of metals (Briggs et. al. 1997); such high concentrations can easily give rise to toxicities which kill plants. Secondly, the soil acidity can affect plant growth due to the change in the mobility of plant nutrients such as calcium, magnesium, phosphorous, molybdenum and boron present in the soil (Kannan, 1991).

Table—5.2: pH regime of the soil types at Kaliveli (Briggs et. al., 1997)

Soil	*pH regime*	*Samples*	*Samples (%)*
Acidic	< 5.5	2, 3, 8, 10, 16, 21	27
Neutral	6-8	1, 4, 5, 6, 7, 9, 11, 12, 13, 14, 15, 11, 18, 19, 20, 21	72
Alkaline	> 8.5	None	-

The acidic soils of Ferdous Nagar (SS8) and Koonimedu (soil sample from Kaliveli lake, SS10) and Vadagaram (soil sample from Kaliveli lake, SS16) could be attributed to the high organic matter and waterlogged conditions. In areas with high rainfall or under waterlogged conditions tend to leach out the basic cations from the soils; moreover, soils developed in these areas have higher concentrations of organic matter, which contain acidic components and residues. Thus these soils tend to have decreased pH and are acidic in nature. The soils with neutral regime of pH (Table 5.2, Fig. 5.3) indicate the presence of free calcium carbonate ($CaCO_3$), and the low availability of phosphorus, manganese, zinc and copper (Briggs et. al., 1997).

Electrical Conductivity

41% of the soils in the Kaliveli catchment fall under the category of low salinity, 23% medium, 36% high salinity (Table 5.3, Fig. 5.2b). The high salinity soils were identified at Aruvadai (agrifield) (Plate 5.1), Koluvari, Kunimedu (Kaliveli lake), and Seyyankuppam (Fig. 5.4). The salinity problem in these areas can be related to the over extended period of evaporation and evapotranspiration of soil exceeding the downward percolation of rainfall, irrigation water, and the 'salt spray' due to the proximity of sea (Todd,1959). To some extent the phenomenon of 'efflorescence' can also held responsible, which is the loss of water by upward movement, where the water tables are closer to the surface (Todd, 1959). This is the clear case with soils present at Pudupattu, Ferdous Nagar, and Kunimedu which showed high salinity levels in the soil.

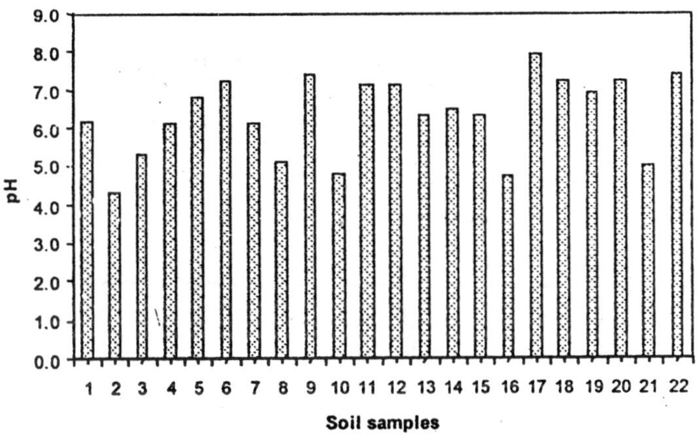

Fig. 5.2a: EC of soil samples collected in-and-around Kaliveli wetland, June 2001.

Nutrient Status

The total nitrogen and total phosphorus content of the soils has been presented as Fig. 5.5 a, b and Table 5.1. In general, the phosphorus levels appear to be negligible when compared to the nitrogen. The relatively high total nitrogen content in some of the soils could be due to:

Plate 5.1: One of the soil sampling sites at Aruvadi; the white sandy saline soils are common along the Yedayanthitu estuary.

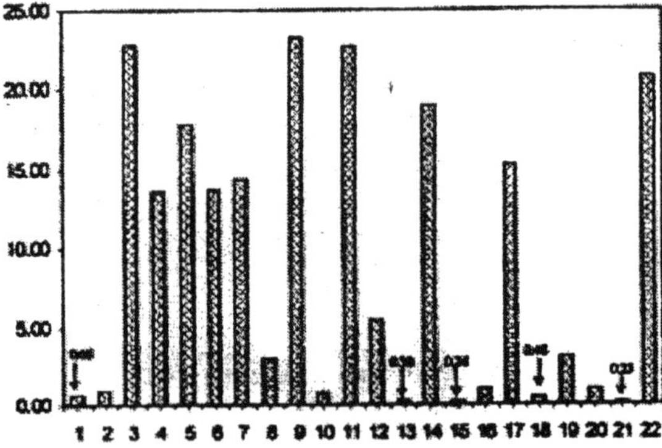

Fig. 5.2b: EC of soil samples collected in-and-around Kaliveli wetland, June 2001.

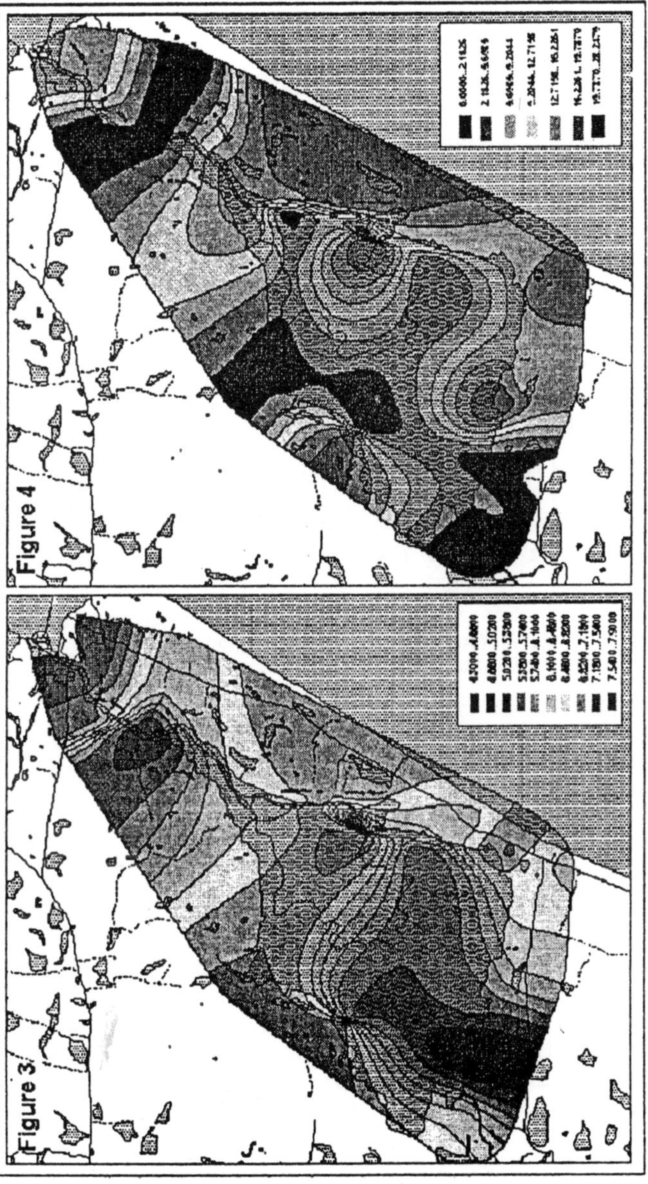

Contours of pH (Fig. 5.3) and EC (Fig. 5.4) of the Soils found in and around Kalveli lake

(i) Aruvadai (SS21, SS22) could be because of fertilizer applications (samples were collected from the Kalive..i lake encroached by agrifields).

(ii) Pudupattu (soil sample from pond, SS5), Ferdous Nagar (Kallveli lake soil sample, SS8) and Kunimedu (Kaliveli lake soil sample SS9) could be attributed to the mineralization of organic matter due to wet soils.

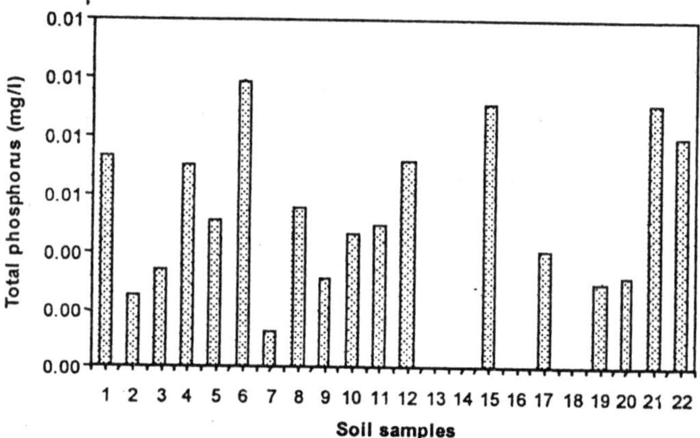

Fig. 5.5a: Total phosphorus of soil samples collected in-and-around Kaliveli wetland, June 2001.

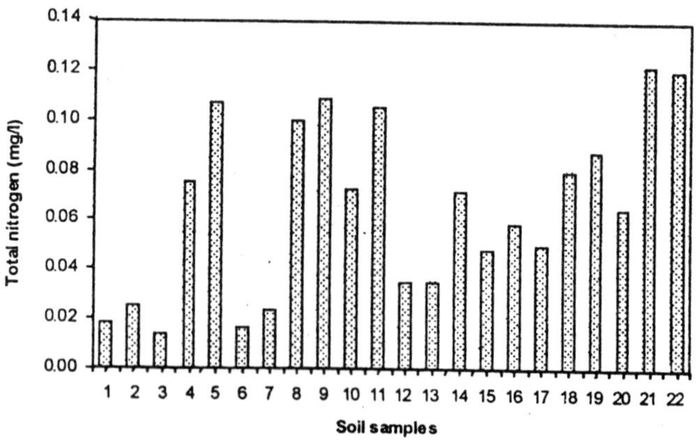

Fig. 5.5b: Total nitrogen of soil samples collected in-and-around Kaliveli wetland, June 2001

Table—5.3: Soil types based on EC (Briggs et. al., 1997, Garg et al., 1980)

Types of soil	*Samples*	*Samples (%)*	*Use in irrigation*
Low salinity (100-250 umhos)	1, 2, 10, 13, 15, 16, 18, 20, 21	41	Can be used for irrigation for almost all crops and for almost all kinds of soils. Very little salinity may develop which may require slight leaching; but it is permissible under normal irrigation practices except in soils of low permeabilities.
Medium salinity (250-750 umhos)	8, 12, 14, 17, 19	23	Can be used, if a moderate amount of leaching occurs. Normal salt-tolerant plants can be grown without much salinity control.
High salinity (750-2250umhos)	3, 4, 5, 6, 7, 9, 11, 12	36	Cannot be used on soils with restricted drainage. Special precautions and measures are undertaken for salinity control and only high-salt tolerant plants can be grown.
Very high salinity (>2250 umhos)	Nil	Nil	Generally not suitable for irrigation.

REFERENCES

APHA, 1998. *Standard Methods for the Examination of Water and Waste Water*, 20th Edition, Washington DC.

Brusseau M.L. and Bohn H.L., 1996. *Chemical Processes Affecting Contaminant Fate and Transport in Soil and Water, In Pollution Science*, Pepper. I.L., Gersa C.P, and Brusseau M.L. (eds.), Academic Press, New York, pp. 63-75.

Briggs. D, Smithson. P, Addison K. and Atkinson K, 1997. *Fundamentals of the Physical Enviornment*, Routledge, London, pp: 334-337.

Garg S. K. 1980. *Irrigation Engineering and Hydraulic Structures*, Khanna Publishers, Delhi, p: 1150.

Kannan K, 1991, Fundamentals of Environmental Pollution, S. Chand and Company, India, p: 336.

Okalebo 1993. *Laboratory Methods of Soil and Plant Analysis Soil*. Science Society of East Africa Technical Publication, Kenya.

Todd D.K, 1959. *Ground Water Hydrology*, John Wiley & Sons, New York, p: 493.

SECTION—IV

WATER QUALITY OF KALIVELI LAKE

6

Water Quality of Kaliveli Lake

Abstract

The aim of this chapter is to elucidate the interrelationships of various water quality variables, identify the pollutants, if any, and to determine the trophic status of the lake. The dynamics of lake water quality were discussed over time and space. Where possible, efforts were made to identify the likely source of these pollutants. Also, the water quality of the lake was assessed for the criteria of BIS (1994) for fish and aquaculture.

Introduction

Lakes are a product of their catchments; therefore, a lake's water quality reflects the condition and management of the lake's catchment. Thus, monitoring water quality of a lake is essential to determine the source of potential or actual lake impairment and to provide a basis for selecting appropriate restoration and protection techniques.

Kaliveli, apart from being a seasonal wetland, the dynamic equilibrium between the fresh water and seawater makes Kaliveli water quality more interesting. In this chapter, a detailed study on the water quality of Kaliveli has been presented.

Materials and Methods

Materials

The materials used for various analytical purposes were as follows:

Chemicals

All the chemicals used was AR-grade and doubly distilled water was used for all the analytical work. The reagents were prepared all fresh when required and standardized, except for $SnCl_4$, which has a consistency of over 6 months (APHA, 1998).

Equipment

(i) pH EP-pH electronic paper, Hanna instruments, range: 0.0-14.0 pH, resolution: 0.1 pH, accuracy: ± (0.2 pH

(ii) Systronics Griph 'D' pH meter 327, range: 0-14 pH, relative accuracy: ± 0.05 pH, readability: ± 0.01pH.

(iii) Naina Digital Field Conductivity meter, NDC 730, range 100 u mhos-100 m mhos in 3 ranges, automatic temperature compensation from 10° C to 60° C.

(iv) Systronics Dissolved Oxygen meter, a battery operated portable instrument for the estimation of DO and temperature. Measurement of DO is indicated in ppm and temperature in °C. The DO ranges are automatically temperature compensated for the solubility of O_2 in water and permeability of the probe membrane. Salinity compensation is manual. Probe is a clark type membrane covered Polarographic Sensor. DO accuracy: ± 1% of full scale at calibration temperature or 0.1 ppm, whichever is higher Temp Accuracy: ± 0.5 C

(v) Anamed Top Pan electronic balance, MX 7301 A, Capacity: 100 gms, resolution: (0.001 gms.

(vi) Roy :op pan balance, KTP-200CSI, Capacity: 2 Kg, Sensitivity: 0.1 Gm.

(vii) Systronics UV-visible spectrophotometer 108, photometric range : accuracy ± 0.005 Abs at 1 Abs, repeatability ± 0.002 abs at 1 Abs, Cuvettes : 10mm.

(viii) Gambaks BOD incubator, G.E 105, range: 5-50°C, Digital temperature controller, temperature uniformity : ± 0.5°C

(ix) Hemco over, Range: 30°C- 50°C.

(x) Tarsons vaccum filter, 250 ml capacity

(xi) Van-Dorn sampler,

(xii) Seechi disk: self-made.

Glassware and plastic ware: All vensil and borosil glassware was used throughout and, PVC temperature and alkaline resistant plastic ware was used for sample collection.

Methods

Water Quality Monitoring: Sampling was done on monthly basis from Nov '98 (summer). Subsequently, sampling was done from Feb '01 until Oct '01 during alternate months for corroborating the consistency of data.

Samples were collected from 5 sampling stations (Fig. 6.1), labelled SW1-SW5, distributed between Kunimedu (the fresh water part of Kaliveli) and Erapanni (the estuarine part of Kaliveli).

Kaliveli holds water only for 5-6 months, during October till March. Later the lake dries up very rapidly. So as to monitor the change in water quality in the watershed, two surface water samples were collected from the ponds adjacent to Kaliveli lake, PD1 (Ferdous Nagar, near fresh water part of Kaliveli) and PD2 (Vandipalayam, near the estuarine part of Kaliveli) (Fig. 6.1).

Kaliveli benefits from both southwest (June-September) and northeast (October-December) monsoons. Thus, the water quality results had been discussed considering the period, June-Dec as monsoon, and Jan-May as post monsoon. The rainfall at Kaliveli has been presented as Fig. 6.2.

The distance between the sampling stations was calculated based on the shortest distance that ' a fish would swim', rather than 'how a crow would fly'. In other words, the shortest distance between the two sampling stations along the lake was considered (Table 6.2).

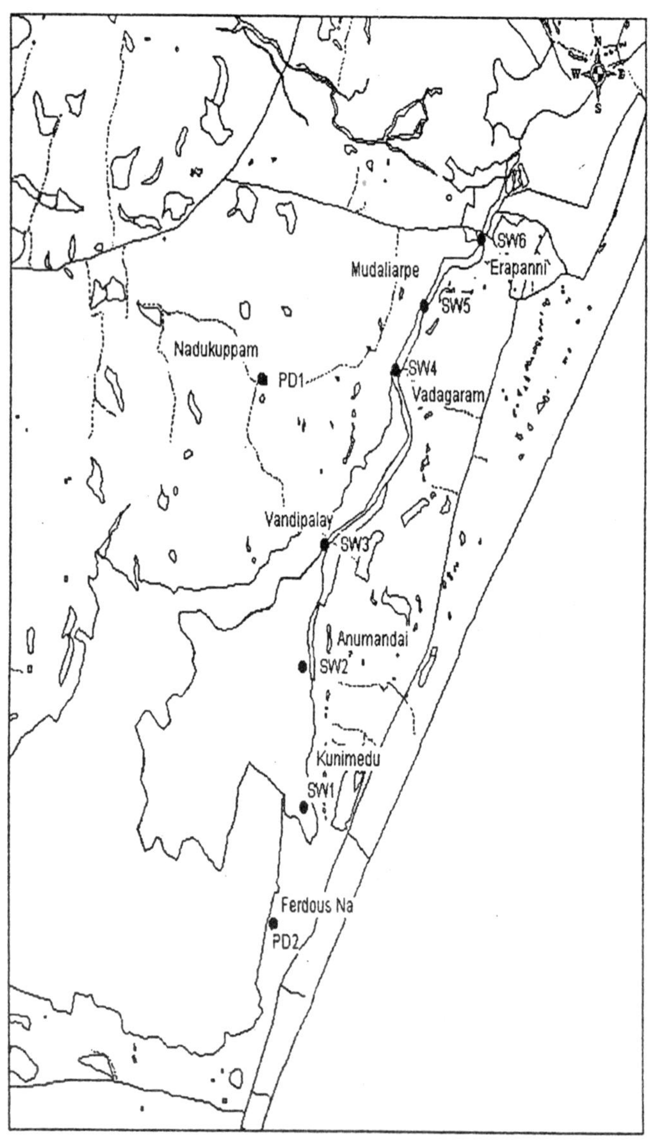

Fig. 6.1: Location of surface water quality sampling stations

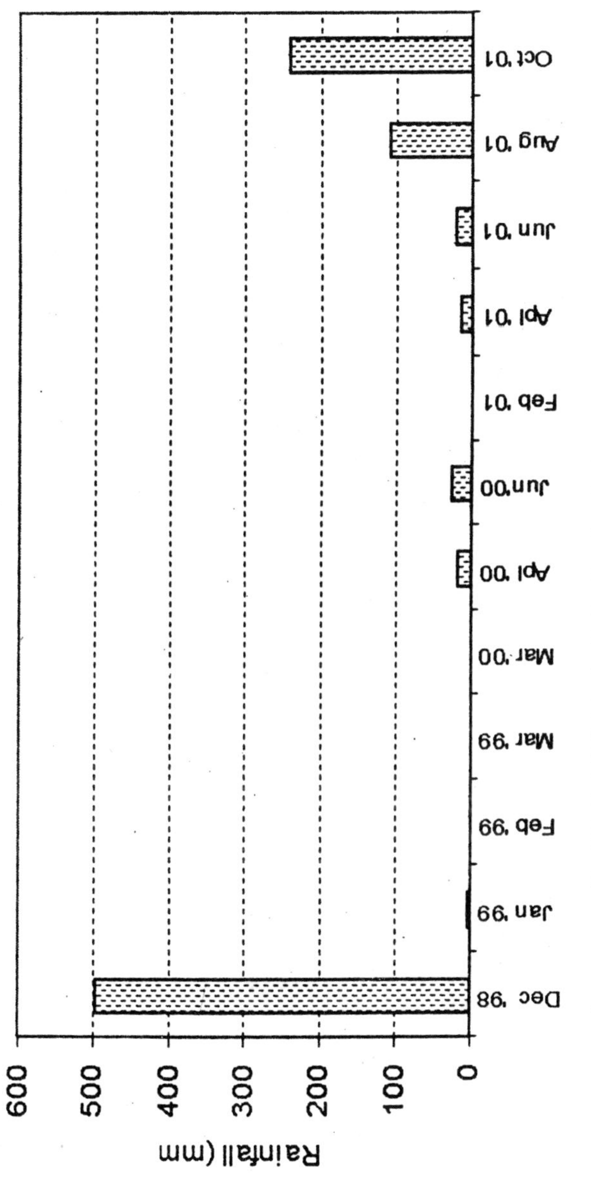

Fig. 6.2: Rainfall data for Kaliveli region during the study period

Water quality parameters: The analytical procedures adopted for the analyses of water quality were as detailed in the standard methods for examination of water and waste water (APHA, 1998).

The following methods were used for estimating the various water quality parameters.

Estimating pH

pH was estimated potentiometrically using pH meter which was calibrated with freshly prepared buffer solutions. As pH is a function of temperature, temperature of the samples was also recorded. The samples were stirred while recording the results for maintaining the homogeneity of the contents.

Estimation of Electrical Conductivity (EC)

Electrical conductivity was estimated using a portable conductivity meter equipped with a dip-type conductivity cell. The EC meter was calibrated with standard KCl solutions as per standard methods (APHA, 1998)

Estimation of Acidity

Acidity was estimated by volumetric method using acid-base titration with NaOH as titrant (APHA, 1998).

Estimation of Alkalinity

The volumetric acid-base titration was used to estimate the alkalinity of the samples. HCl was used as the titrant (APHA, 1998)

Estimation of Hardness

Hardness was estimated by complexometric titration method, where the samples were titrated against di-sodium salts of EDTA with the appropriate metallochromic indicators. While calmagite was used as the indicator for total hardness, ammonium purpurate was used for calcium hardness (APHA, 1998).

Estimation of Chlorides

Chlorides were estimated through argentometric titration of the neutral or near alkaline samples against standard silver nitrate solution (APHA, 1998).

Estimation of Dissolved Oxygen

Dissolved oxygen was estimated both by electrometric and titrimetric (azide modification) method based upon the redox reaction of manganese salt (APHA, 1998).

Estimation of Phosphorus

Phosphorus was determined as phosphates (PO_4) by the stannous chloride method. Different forms of phosphorus were analyzed by filtering the samples through a 0.45 *u* membrane filter and subjecting the samples to selective acid digestion.

Use of detergents was avoided for cleaning the glassware as detergents are known to contain substantial amounts of phosphorus. Instead, the glassware was rinsed with warm HCl. Also, the membrane filters were soaked for more than 24 hours in 500 ml distilled water prior to filtration of the samples. Further, the contribution of PO_4 by the membrane filters to the samples was checked by adding a blank of distilled water filtrate (APHA, 4500- P.D)

Estimation of Nitrogen

Organic and ammonical nitrogen were determined by macro-kjeldahl method. Ammonical nitrogen was determined during the preliminary distillation and the samples were further digested for organic nitrogen. Amino nitrogen, ammonia, ammonium nitrogen of many organic materials is converted to Ammonium sulphate in the presence of sulphuric acid. Mercuric sulphate was then digested forming an ammonium complex and then decomposed by Sodium-thio-sulphate. After the decomposition, ammonia was distilled from the alkaline medium and absorbed by indicator boric acid which was then titrated against 0.002N H_2SO_4 to determine the organic-nitrogen (APHA, 4500. NH_3.E). Nitrite-Nitrogen was estimated by calorimetric method through formation of a reddish purple azo-dye produced at pH 2.0 to 2.5 by coupling diazotization of sulfanilamide with N-(1-naphthyl) ethylene diamine dihydrochloride (NED dihydrochloride) (APHA, 4500-NO_2.E).

Nitrate-nitrogen was determined by phenol-di-sulfonic acid method.

The lake is used for fishing alone and not for drinking and irrigation purposes. Thus, the water quality results of Kaliveli were compared with the water quality criteria of BIS for fresh water fish culture (1994) alone.

The data of Kaliveli water quality has been presented as Annexure at the end of this chapter.

Results and Discussion

pH

pH of Kaliveli lake, during the monsoon period, varied between 6.7 (SW2, Dec '98) and 9.7 (SW2, Oct '01) (Fig. 6.3a). The spatial variance of pH along the lake was not very significant. Temporarily, pH regime of the lake tends to shift from alkaline to mildly acidic by the end of peak monsoons. This could be attributed to the fact that precipitation, which is mildly acidic due to the dissolution of atmospheric CO_2, could have contributed to the change in pH.

During the post-monsoon period, pH of the lake varied between 5.4 (SW3, March 2000) and 8.9 (at SW2 during Jan '99, March '00, and April '00) (Fig. 6.3b). Spatially, pH of the lake did not vary significantly in any given month (Fig. 6.4). However, temporarily the pH fluctuated widely at site SW3. This could be due to the fact that at this point the lake becomes narrow and shallow, and as the organic matter decomposes during the post monsoon, pH tends to decrease.

During monsoon, nearly 94% samples of Kaliveli lake were within the stipulated limits of the fish culture of BIS (1994); of these, 20% of the samples fell either below or above the permissible limits during the post monsoons.

A majority of the samples of Kaliveli were within the permissible limits of irrigation water criteria of BIS (1994) for pH. A total of 97% samples were within the permissible limits during monsoon and 91% during post monsoon. Only 5.7% of the samples fell below or above the permissible limits of pH during the post monsoons.

Alkalinity

The alkalinity of Kaliveli lake during the monsoon period varied between 220 mg l^{-1} (SW6, June '01) and 6 mgl-1 (SW6, Oct

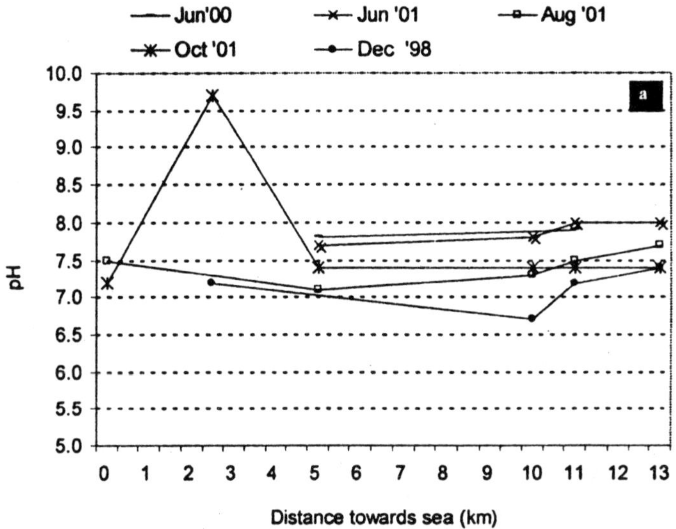

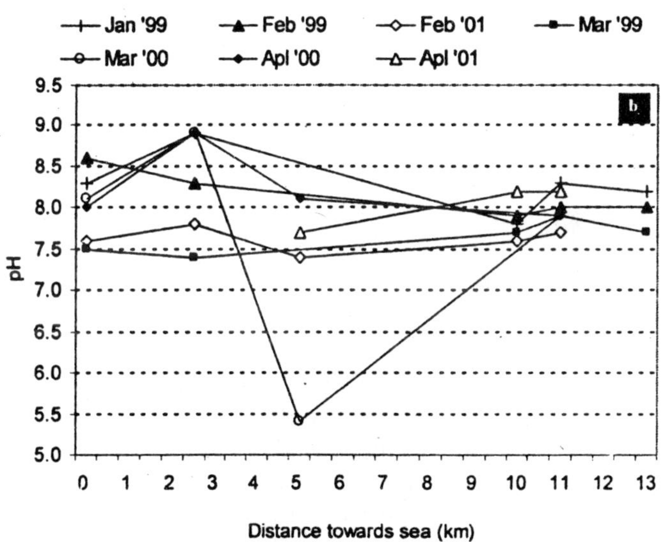

Fig. 6.3: pH of Kaliveli lake during (a) monsoon and (b) post monsoon

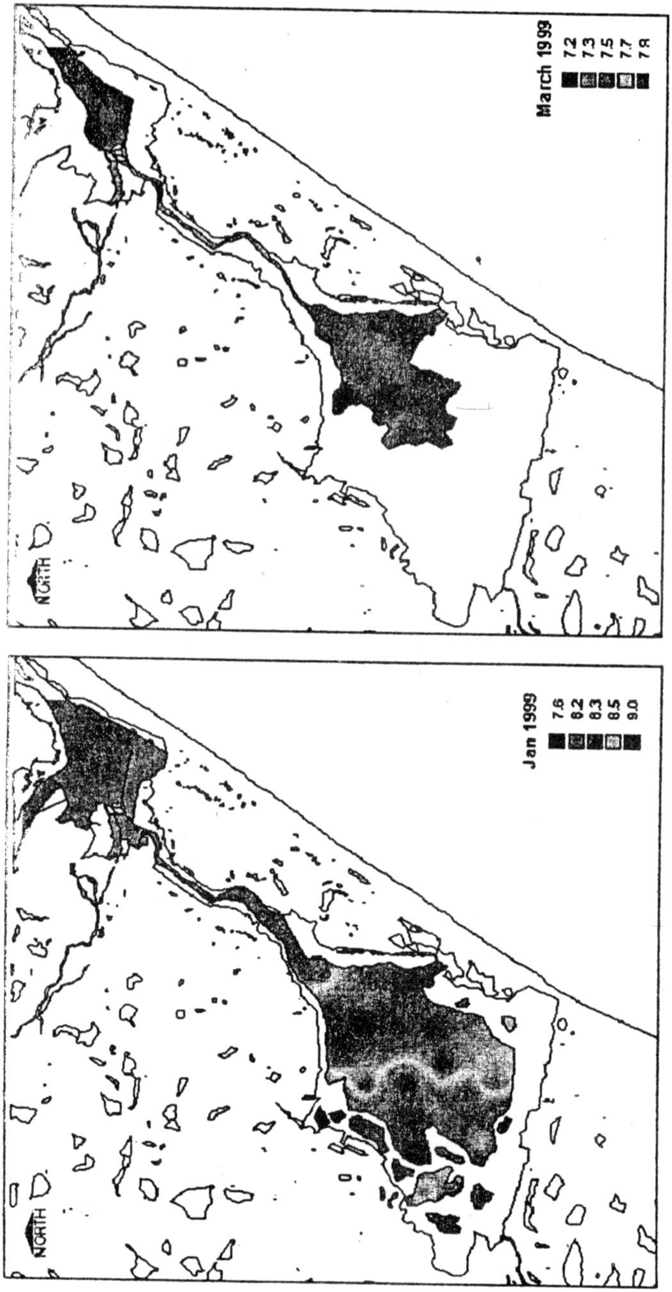

Fig. 6.4: The spatial pattern of pH immediately after the monsoons, Jan 1999 and during the summer March 1999

'01) (Fig. 6.5a). During the post monsoon the alkalinity varied between 20 mg l^{-1} (SW1, 4 and 5, Feb '01) and 268 mg l^{-1} (SW5, Mar '99) (Fig. 6.5b).

The alkalinity values partially correlated positively with pH. The pattern suggests that the dilution factor due to monsoon makes the lake mildly acidic; and the subsequent drying up of the lake and the influx of sea water makes the lake alkaline. This is further corroborated by the increase in alkalinity during the peak summer. Also, the alkalinity of the lake was found to increase as the lake approaches sea.

Only 50% of the samples were within the permissible limits of water quality criteria of BIS (1994) for fresh water fish culture during the monsoon. This is due to the factor that the lake becomes acidic during the monsoons. As the alkalinity levels in the lake rise during the period of post monsoon, 45% of the total samples fell within the stipulated alkalinity levels of BIS (1994) for fresh water fish culture.

Acidity

During the late months of monsoon the acidity values were the highest, reaching 210 mg l^{-1} (SW3, Oct '01) (Fig. 6.6a). The rise in the acidity levels could be attributed to the precipitation and storm run-off, which are mildly acidic.

During the months of post monsoon the acidity levels ranged between 80 mg l^{-1} (SW1, Feb '01) and as low as 0 mg l^{-1} (all sites along the lake, throughout post monsoon except Feb '01) (Fig. 6.6b).

The acidity levels of the estuarine part of Kaliveli fluctuated between 0 mg l^{-1} (all sites except Feb'01) and 60 mg l^{-1} (SW4,SW5,Feb'01) (Fig. 6.6b).

Electrical Conductivity

The EC of Kaliveli, during the monsoon varied between 0.083 (SW4, Dec '98) and 41 (SW4, Jun '01) (Fig. 6.7a). Due to dilution factor of precipitation the EC decreases rapidly as the monsoon pickup (Fig. 6.7b). The EC which usually would be 40 mmhos in the month of June, by the end of Dec the EC decreases as low as 40 times.

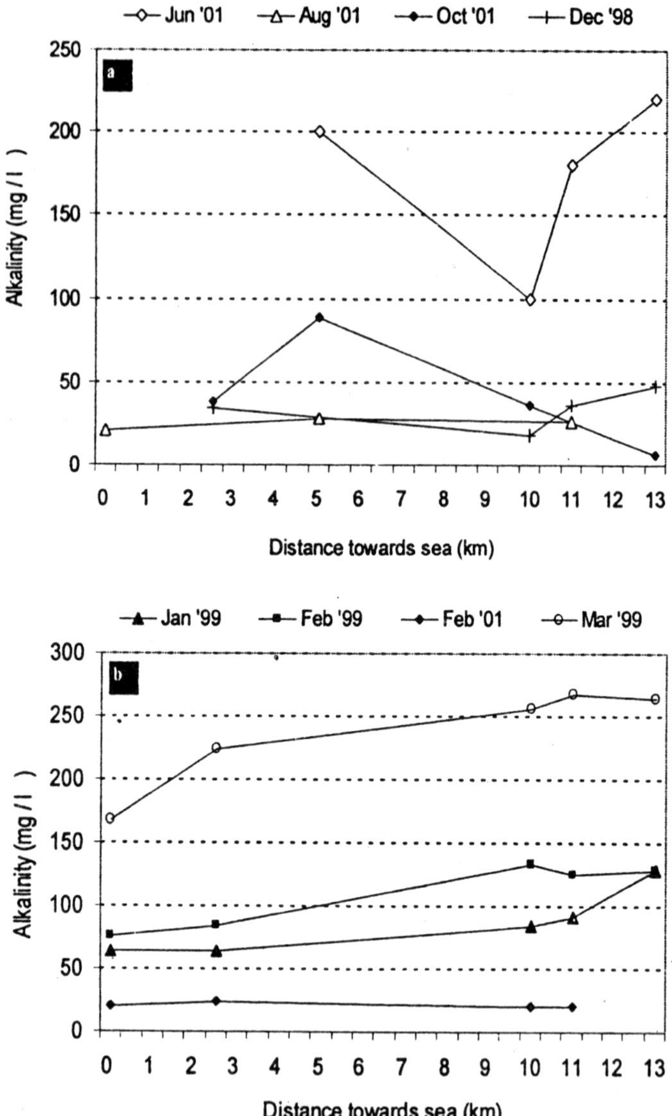

Fig. 6.5: Alkalinity of Kaliveli lake during (a) monsoon, and (b) post monsoon

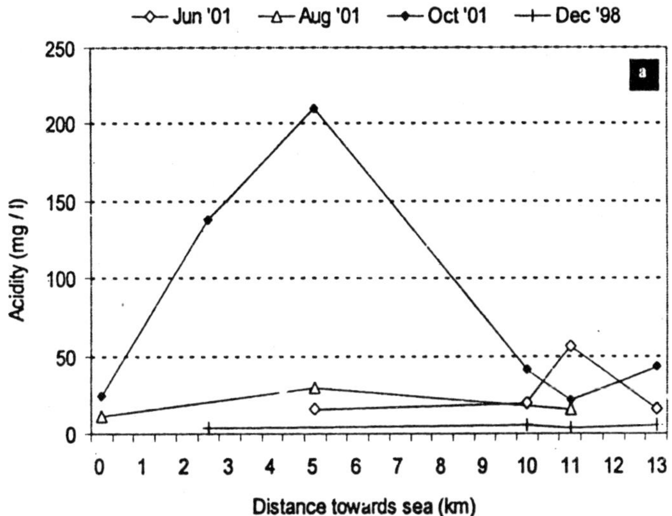

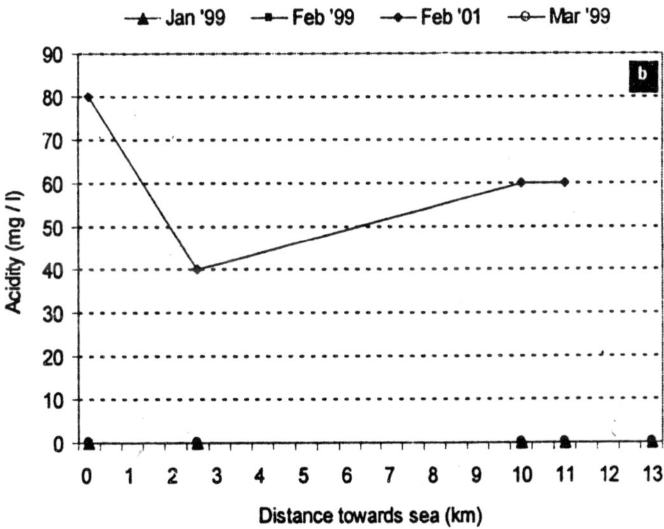

Fig. 6.6: Acidity levels of Kaliveli during (a) monsoon and (b) post monsoon

Once the monsoon wanes away the EC keeps rising again due to evaporation losses and shrinking of the water spread. At site SW1 the EC fluctuates between 1.63 and 12.85 mmhos. As one approaches the sea the amplitude of EC change is much more dramatic. In the estuarine part of Kaliveli, Yedayanthitu estuary, the EC fluctuates anywhere between 0.16 and 53 mmhos.

The spatial pattern of the EC across the lake has been illustrated as Fig. 6.8.

The salinity condition of Kaliveli, as reflected by EC, is governed by the factors such as local precipitation, influx of surface runoff, proportionate admixing of sea water with fresh water and evaporation.

Chlorides

The chloride levels of Kaliveli fluctuated in consonance with that of EC. The peak levels of chlorides were observed in the beginning of the monsoon, 31740 mg l^{-1} (SW6, Jun '01); and the lowest, 6 mg l^{-1} (SW3, Dec'98), by the end of the monsoon (Fig. 6.9a). Thus, the dilution factor due to precipitation plays a crucial role in the drastic decrease in the chloride levels during the monsoon.

During the post monsoon period, chloride levels ranged between 281 mg l^{-1} (SW1, Jan '99) and 13042 mg l^{-1} (SW6, Mar '99) (Fig. 6.9b). By the end of the post monsoon, the chloride levels rise by as much as 12 folds at Kunimedu, and 30 folds at Erapanni.

Total Hardness

During the monsoon total hardness levels of Kaliveli ranged between 24 mg l^{-1} (SW4, Dec '98) and 11000 mg l^{-1} (SW6, Jun '01) (Fig. 6.10a). The beginning months of Kaliveli had the hardness levels close to 11000 mg l^{-1} (SW6, Jun '01) and by the end of the monsoon the hardness levels dropped as low as 24 mg l^{-1} (SW4, Dec '98).

During the post monsoon, the total hardness levels reached as high as 4540 mg l^{-1} by March 1999 at SW6 (Fig. 6.10b). In the fresh water part of Kaliveli the total hardness levels ranged between 128 mg l^{-1} (SW1, Jan 1999) and 3274 mg l^{-1} (SW2, Feb '99). As expected the total hardness levels increase as the lake approaches the sea.

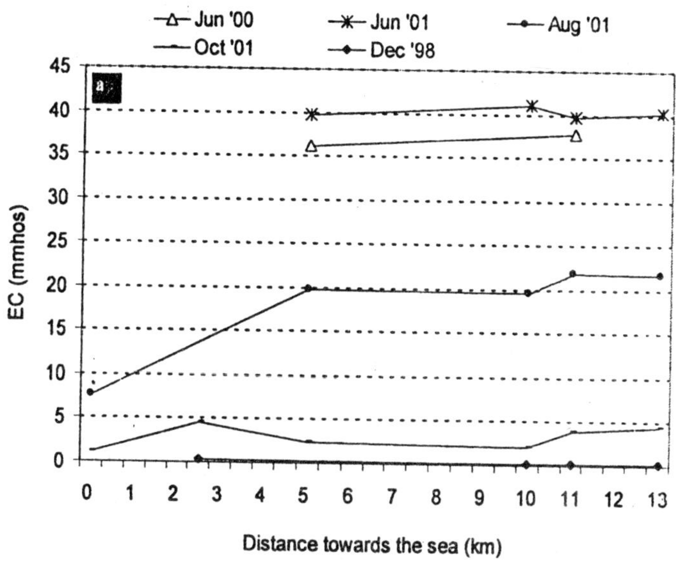

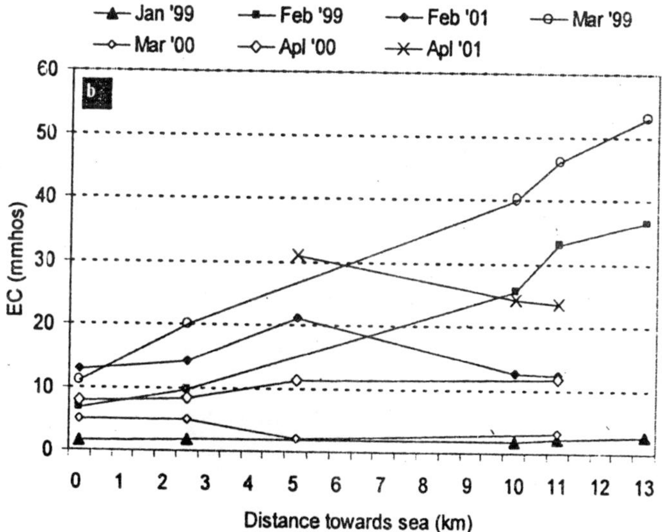

Fig. 6.7: Electrical conductivity of Kaliveli lake during (a) monsoon, and (b) post monsoon

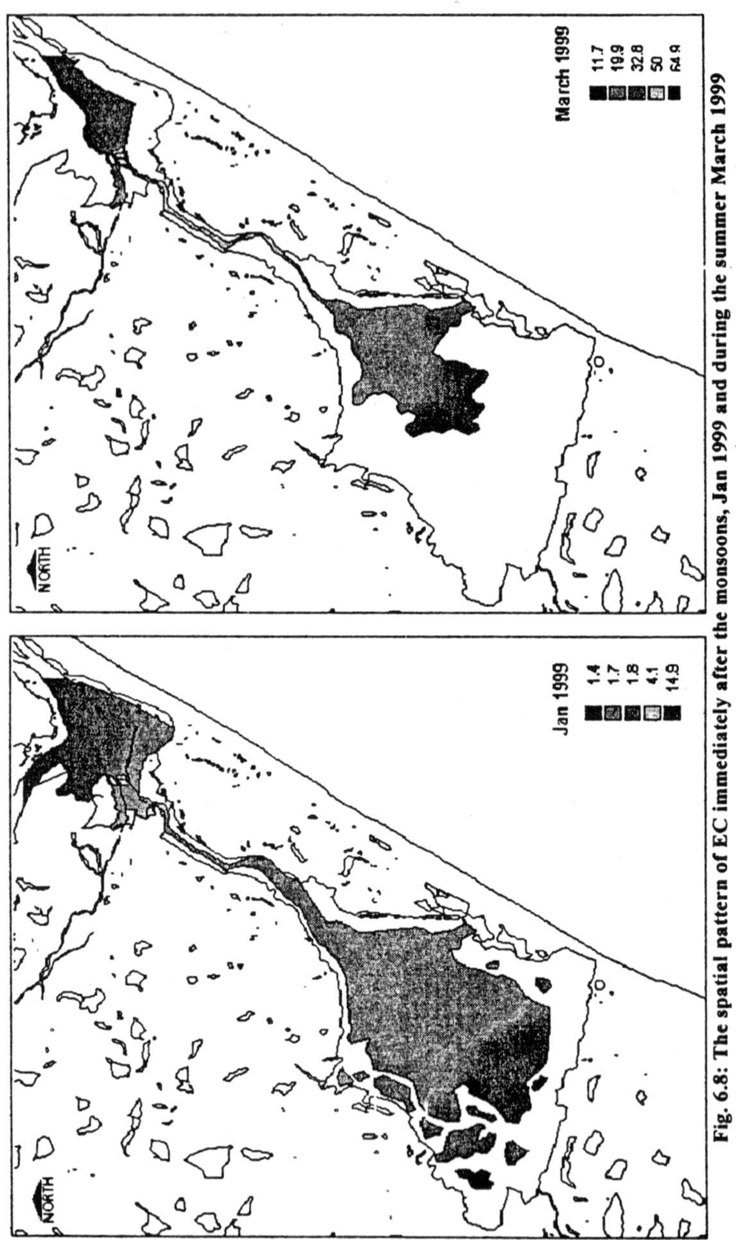

Fig. 6.8: The spatial pattern of EC immediately after the monsoons, Jan 1999 and during the summer March 1999

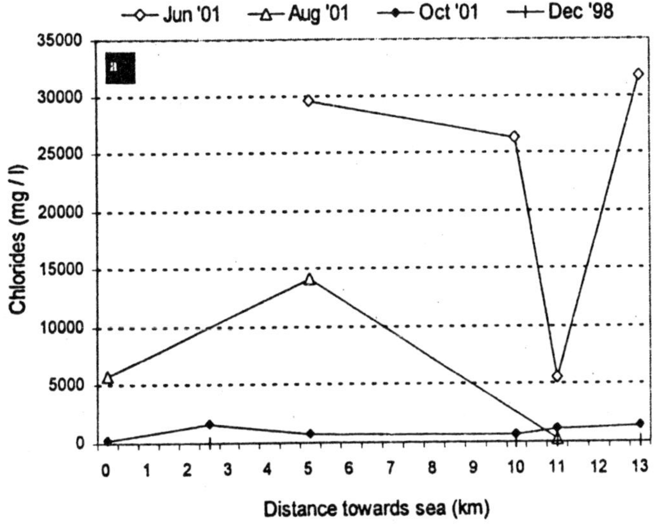

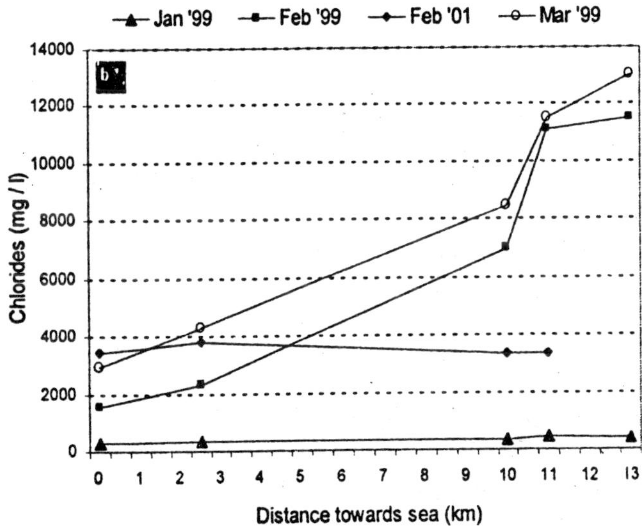

Fig. 6.9 Chloride levels of Kaliveli lake during (a) monsoon, and (b) post monsoon

Thus, seawater forms the chief and predominant source of hardness in Kaliveli.

Calcium Hardness

The pattern of calcium hardness followed closely with that of total hardness. Calcium hardness levels in the lake during the monsoon ranged between 10 mg l^{-1} (SW4, Dec '98) and 1800 mg l^{-1} (SW6, Jun'01) (Fig. 6.11a). The calcium hardness levels decreased as the monsoon progressed. This could be attributed to the dilution factor due to precipitation.

During the period of the post monsoon, calcium hardness of Kaliveli ranged between 19 mg l^{-1} (SW1, Feb '01) and 810 mg l^{-1} (SW6, Mar '99) (Fig. 6.11b). The peak levels of calcium hardness were observed towards the end of the post monsoon. Spatially, the amplitude of fluctuation was higher as the lake approaches sea.

Sulfates

In general, sulfate levels in the lake decreased as the lake approached sea, except during Aug '01. During the monsoon period the sulfate levels were the lowest at SW4 and SW6 (1.4 mg l^{-1}, Dec '98) (Fig. 6.12a), and the highest at SW5, (526 mg l^{-1}, Aug '01). The concentration of sulfates at the beginning of the lake and in the estuary was same, taking an exception of the site SW3.

During the months of post monsoon, the sulfate levels did not vary significantly across the lake. The highest levels of sulfate were observed at SW2 (Feb '01, 294 mg l^{-1}) and the lowest at SW4 (2.4 mg l^{-1}, Jan '99) (Fig. 6.12b).

Total Iron

During the monsoon, highest levels of iron observed was at SW2 (3.9 mg l^{-1}, Oct '01) and the lowest at SW5, (0.3616 mg l^{-1}, Oct '01) (Fig. 6.13a). During the period of post monsoon the highest levels of iron was observed at SW3 (0.321 mg l^{-1}, Jan '99) and the lowest at SW5 (0.168 mg l^{-1}, Feb '01) (Fig. 6.13b). However, no pattern was observed either spatially or temporally.

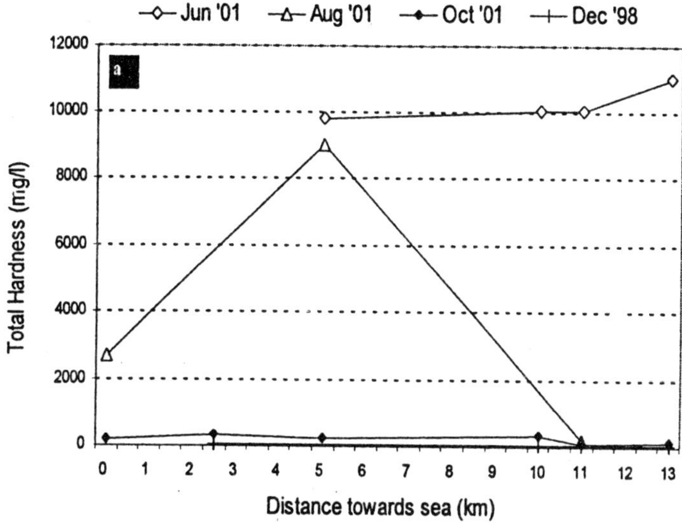

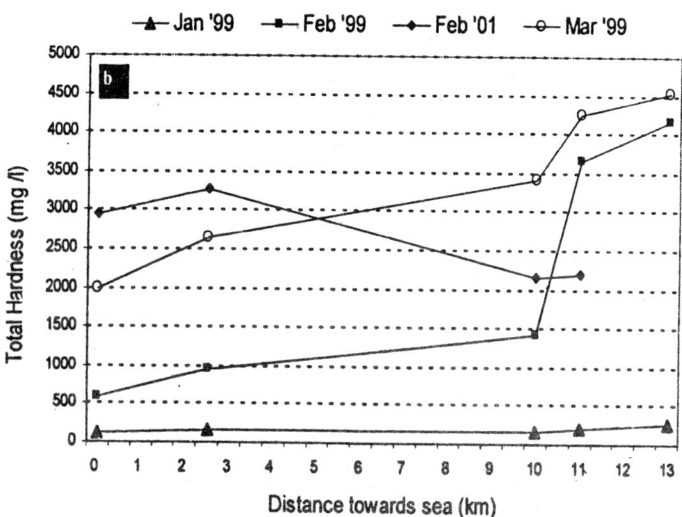

Fig. 6.10: Total hardness levels of Kaliveli lake during (a) monsoon, and (b) post monsoon

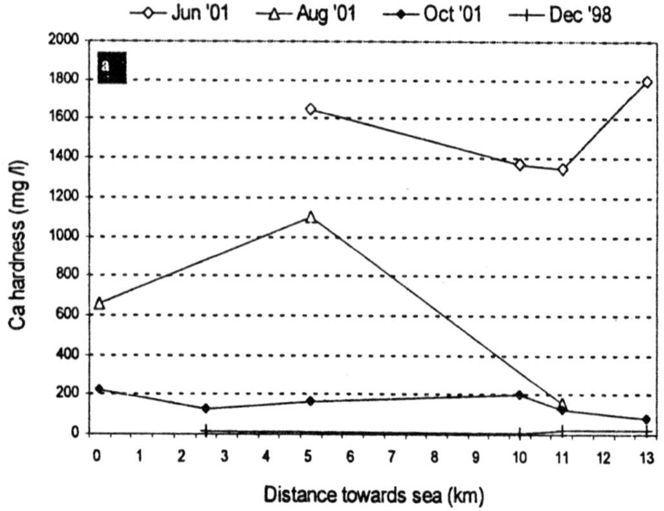

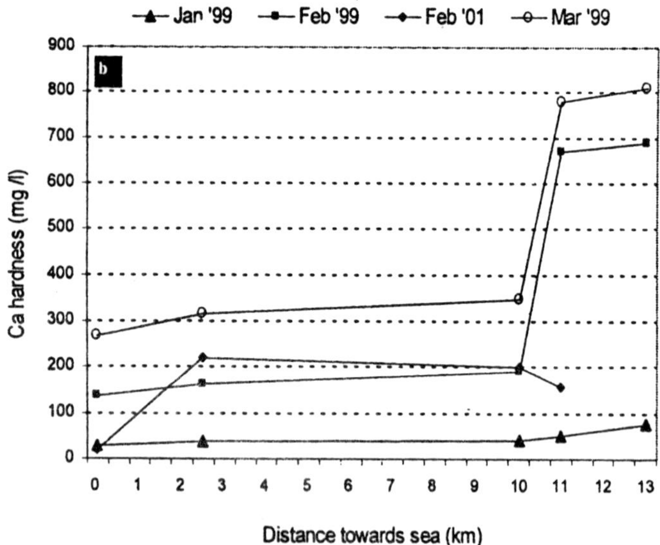

Fig. 6.11: Calcium hardness levels of Kaliveli lake during (a) monsoon, and (b) post monsoon

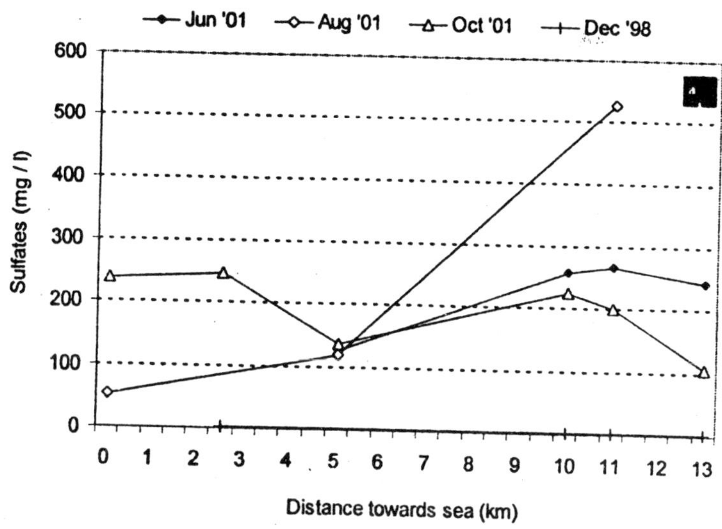

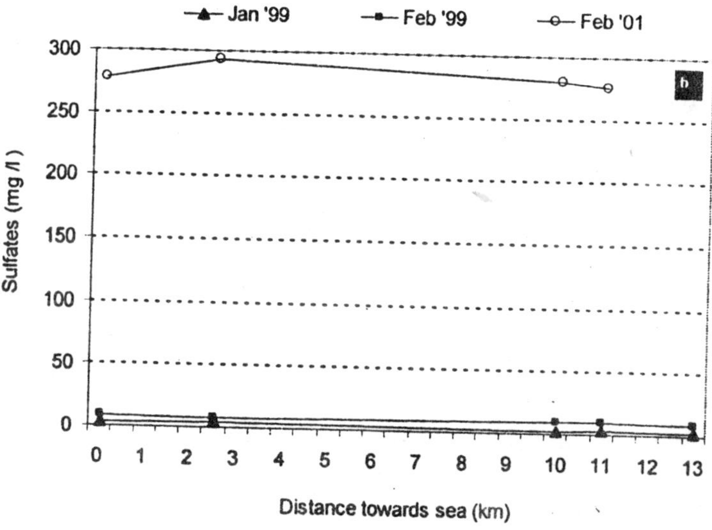

Fig. 6.12: Sulfate levels of Kaliveli lake during (a) monsoon, and (b) post monsoon

The high iron contents during monsoon indicate that surface runoff is the primary source of iron in Kaliveli. The extensive agricultural activities, increased diffusion of ferrous iron from sediment at lower oxygen concentration near sediment surface could be responsible for the high iron levels in the catchment. Similar observations were reported by Sarwar and Rifet (1991).

As per the water quality criterion of BIS for fresh water fish culture (1994) 9% of the samples were within the permissible limits during monsoon and 15% during the post monsoon.

Nitrogen

Nitrites: During the monsoon, highest concentration of nitrites was observed during Dec '98 at SW5 (2.25 mg l^{-1}); while the lowest concentrations were observed at SW5 during Oct '01 (0.0000374 mg l^{-1}) (Fig. 6.14a). Temporarily, the nitrite values were found to be higher during the beginning of monsoons and by the end of the monsoon the concentration of nitrites decrease drastically. Moreover, spatially, the nitrite levels were found to decrease towards the estuary.

During the post monsoon, the highest values were observed at SW4, (0.031 mg l^{-1}, Jan '99), and the lowest at SW5 (0.0000645 mg l^{-1}, Feb '01) (Fig. 6.14b). It was observed that during the beginning of post monsoon, nitrites levels were comparatively higher in the estuary than the fresh water part of the lake. However, during the subsequent months the nitrite levels showed a declining trend towards the estuary.

The least nitrite concentrations observed in the lake can be due to the reduction of nitrites by the phytoplankton such as diatoms (Hutchinson, 1975).

Nitrates: The highest levels of nitrates, during monsoon, were found at SW3 (15.2 mg l^{-1}, Aug '01) and the lowest at SW4 (0.02 mg l^{-1}, Dec '98) (Fig. 6.15a). The lowest levels of nitrogen were observed in the month of Dec '98 due to the dilution factor, as that particular month has received the highest rainfall during the study period.

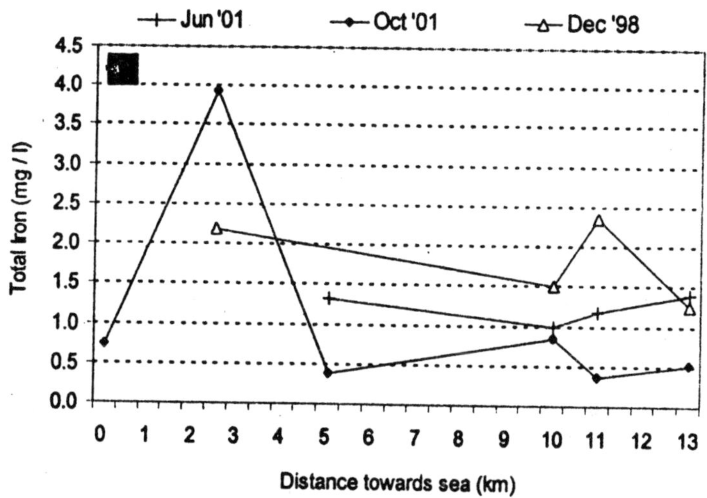

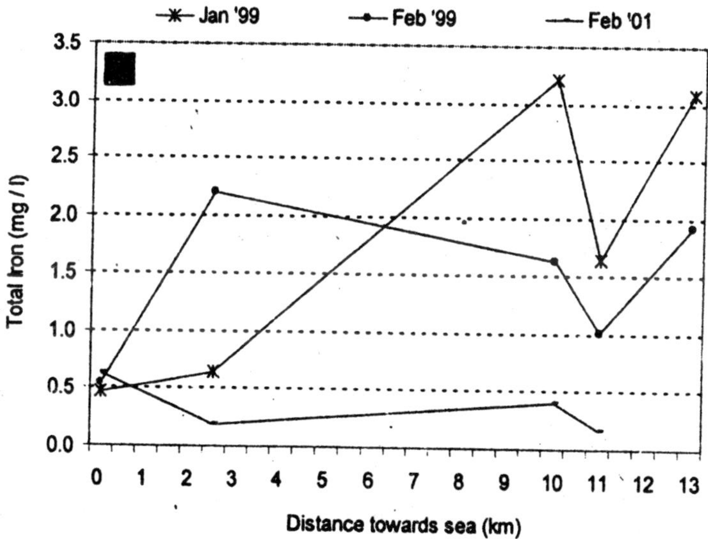

Fig. 6.13: Total iron levels of Kaliveli lake during (a) monsoon, and (b) post monsoon

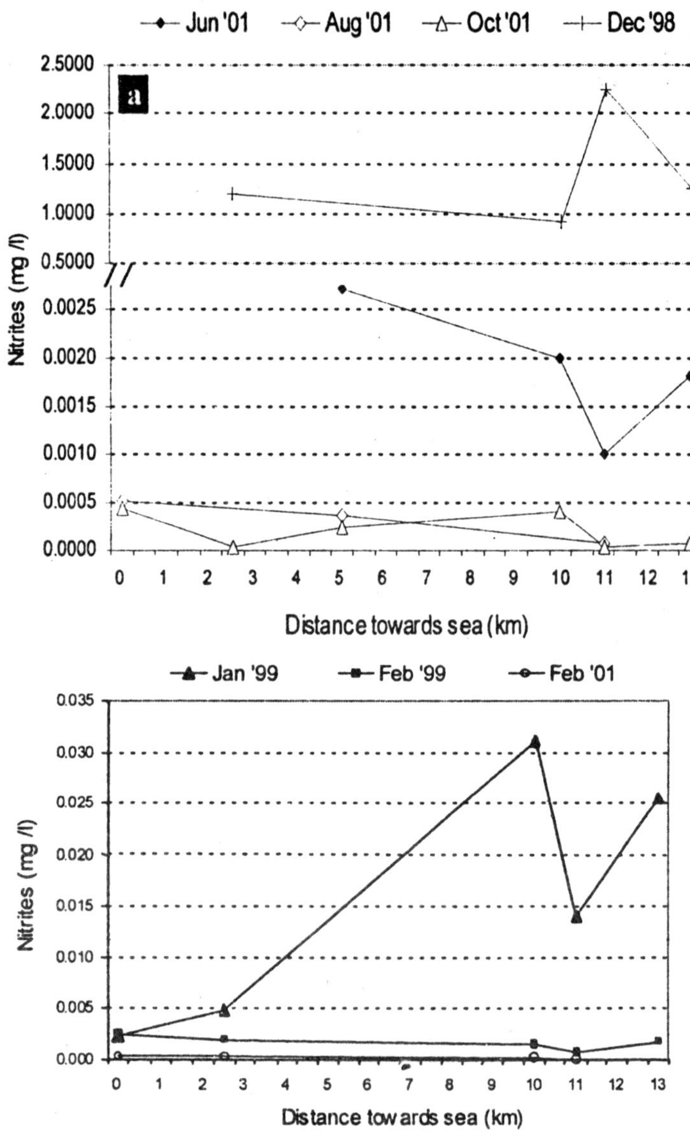

Fig. 6.14: Nitrite levels of Kaliveli lake during (a) monsoon, and (b) post monsoon

During the month of post monsoon the lowest levels of nitrates, 0.095 mg l^{-1} were observed at SW2 (Jan '99), and the highest, 2.509 mg l^{-1} at SW5 (Feb '99) (Fi. 6.15b). The nitrate concentration of the lake tends to increase as the post monsoon picks up.

Ammonical Nitrogen: During monsoon the lowest concentration of ammonia was observed at SW5, 0.0083 mg l^{-1} (Aug '01); the highest concentrations were observed at SW5, 0.43 mg l-1 (Jun '01) (Fig. 6.16a). The amplitude of fluctuation in the ammonia concentration was observed at SW3, and mild fluctuations were observed at SW4.

The lowest concentration of ammonical nitrogen, during post monsoon was observed at SW4, 0.044 mg l^{-1} (Feb '99); and the highest at SW5, 0.535 mg l^{-1} (Feb '01) (Fig. 6.16b). Temporarily the ammonia levels fluctuated widely at SW1 and SW5. Spatially, the ammonia levels tend to decrease till SW4 and again increase.

Organic Nitrogen: The highest level of organic nitrogen during monsoons were observed at SW3 (1.39 mg l^{-1}), Vandipalayam, during the month of Jun '01, and the lowest at SW4 (0.1 mg l^{-1}) (Fig. 6.17a). Vadagaram during the month of Dec '98, in a given month the fresh water part of the lake, till 5 km had more organic nitrogen than the estuarine of the lake. The organic levels were particularly low at site SW4.

During post monsoon, the higher levels of organic nitrogen were observed at SW2, 6.7 mg l^{-1}, during Feb '01; and the lower level at SW1, 0.112 mg l^{-1}, during the month of Jan '99 (Fig. 6.17b). Except at SW2 (Feb '99) the organic nitrogen during the post monsoon were lower than the monsoon period.

Total Nitrogen: During the monsoon period total nitrogen in the lake varied between 0.6709 mg l^{-1} (SW3, Oct '01) and 16.96 mg l^{-1} (SW3, Aug '01) (Fig. 6.18a). The lower levels of total nitrogen observed during the peak monsoon months could be attributed to the dilution factor due to precipitation.

Total nitrogen in the lake, during post monsoon varied between 0.0997 mg l^{-1} (SW2, Jan '99) and 3.97 mg l^{-1} (SW2, Feb '99) (Fig. 6.18b). No specific pattern was observed either spatially or temporarily.

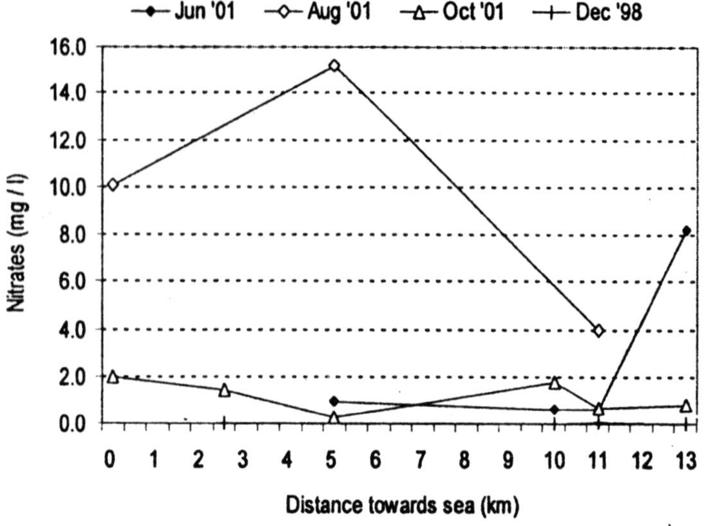

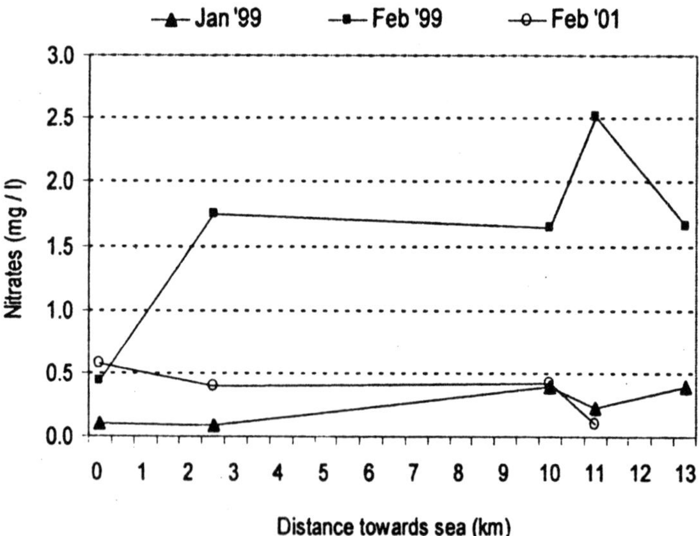

Fig. 6.15: Nitrate levels of Kaliveli lake during (a) monsoon, and (b) post monsoon

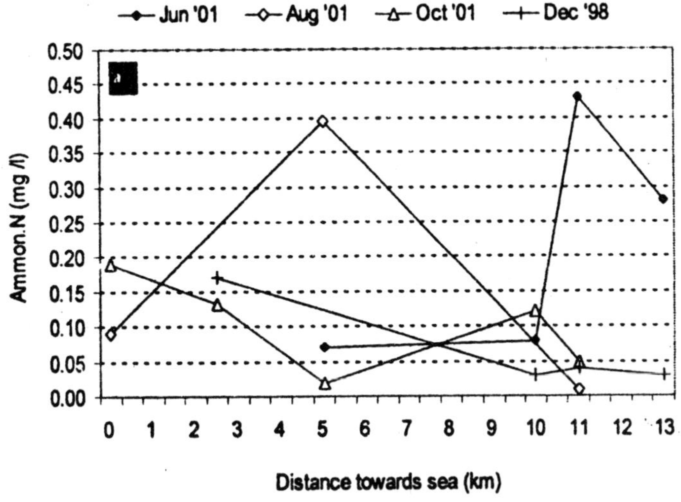

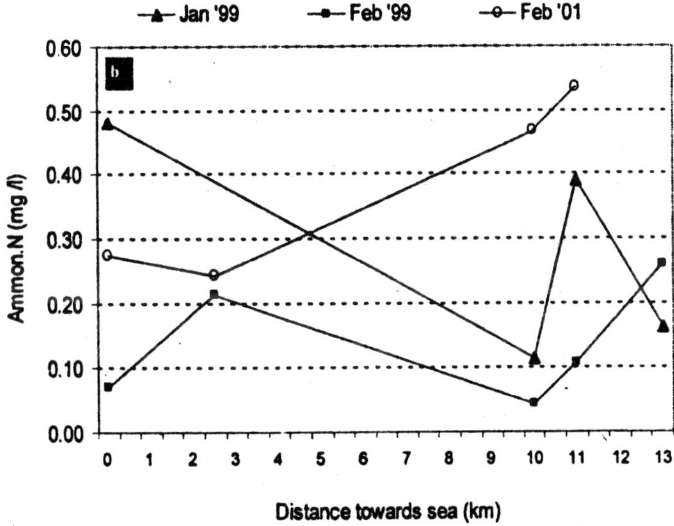

Fig. 6.16: Ammonical nitrogen levels of Kaliveli lake during (a) monsoon, and (b) post monsoon

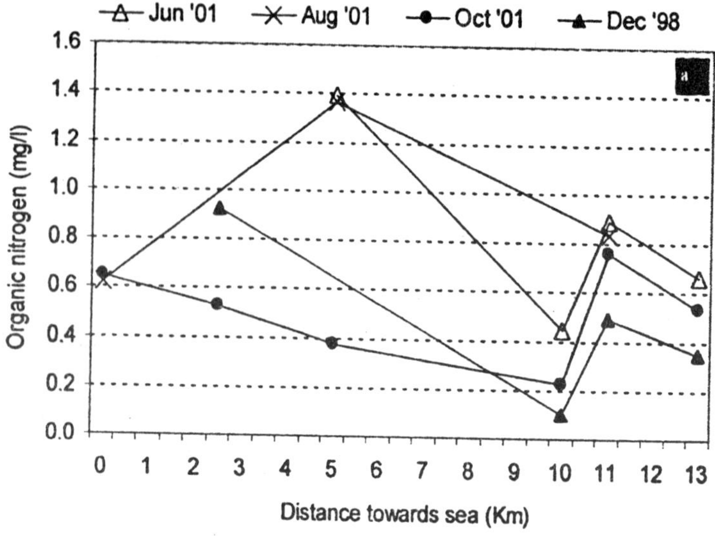

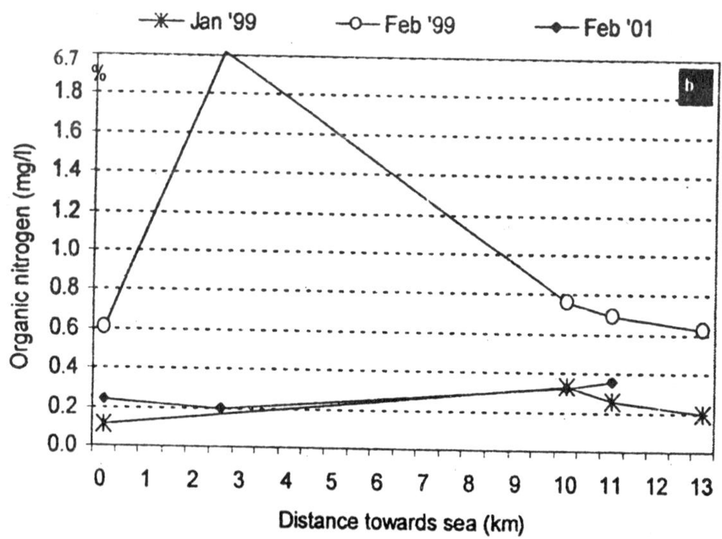

Fig. 6.17: Organic nitrogen levels of Kaliveli lake during (a) monsoon, and (b) post monsoon

Phosphorus

The total phosphorus (TP) levels of Kaliveli during the monsoon period ranged between 0.006 mg l^{-1} (SW5, Aug '01) and 1.427 mg l^{-1} (SW4, Dec '98) (Fig. 6.19a).

During post monsoon, phosphorus levels ranged between 0.03 mg l^{-1} (SW4, Mar '97) and 1.59 mg l^{-1} (SW2, Feb '97) (Fig. 6.19b). It was observed that the fresh water part of the lake had high phosphorus levels when compared to the estuarine part of the lake. Except during the month of Feb '97 the rest of the post monsoon phosphorus levels were found to decrease towards the estuary. Thus, fresh water part of the lake can be considered as more productive than the estuary. One can clearly notice a clear sharp decline in the total phosphorus concentrations at a distance of 11 km from south.

The phosphorus levels tend to increase in the initial period of post monsoon, during the later period a sharp decline was observed in the concentrations of phosphorus. The high TP concentrations were found during the initial months of post monsoon, which suggests that surface runoff is the chief controlling factor of phosphorus in Kaliveli than the sea water. This is contradictory to the observations made by Chandran and Ramamurthi (1984), Upadhyay (1988), and Banerjee and Roy (1966) who have worked on other estuaries of India. Apart from surface runoff, animal vectors also are responsible for the nutrient loads through their excrements. The vast stretch of Kaliveli, till 7 kms towards sea, is essentially a flat terrain supporting various grasses and reeds. The domestic livestock feed upon these grasses, and the excrement droppings add nutrients and organic matter which enrich the lake further (Plate 6.1 and 6.2).

The sharp decline in TP contents can be attributed to the fast depletion of phosphorus by the aquatic macrophytes and reeds, which grow and very rapidly once the lake receives copious rainfall.

According to Nurnberg (1996), lakes with TP more than 0.01 mgl^{-1} are hyper-eutrophic; and as the TP concentration of Kaliveli often exceed 0.01 mgl^{-1}, the lake can be considered as hyper-eutrophic.

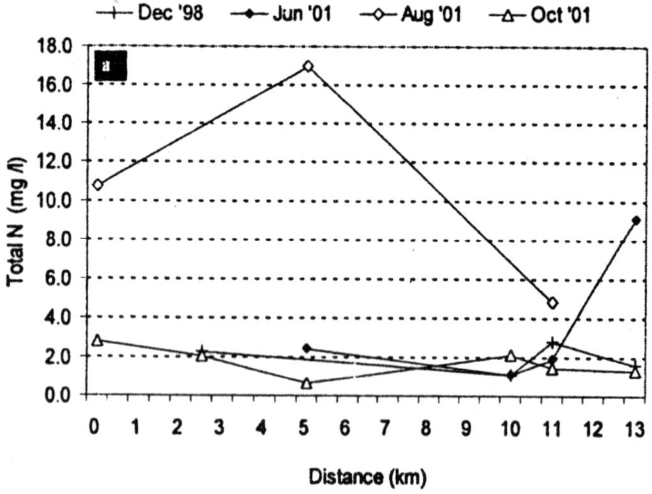

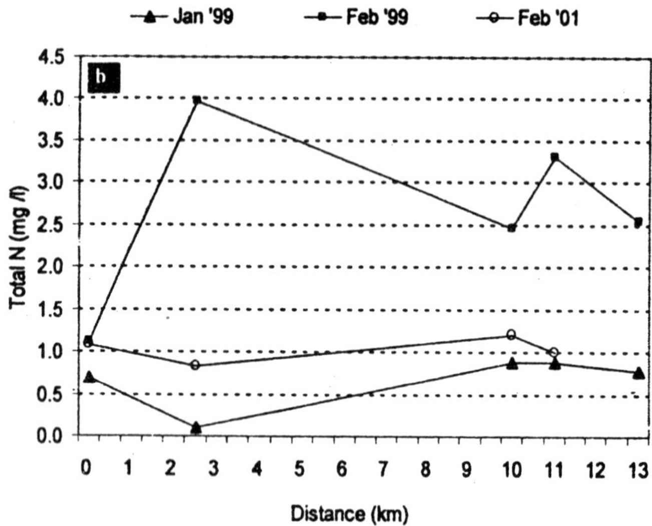

Fig. 6.18: Total nitrogen levels of Kaliveli lake during (a) monsoon, and (b) post monsoon

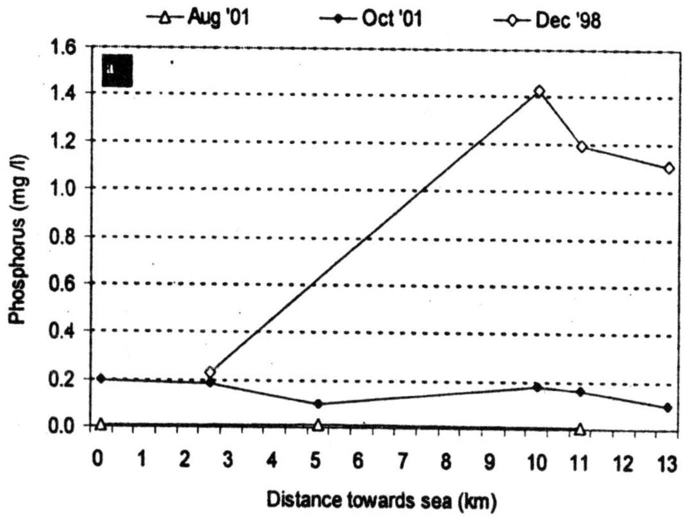

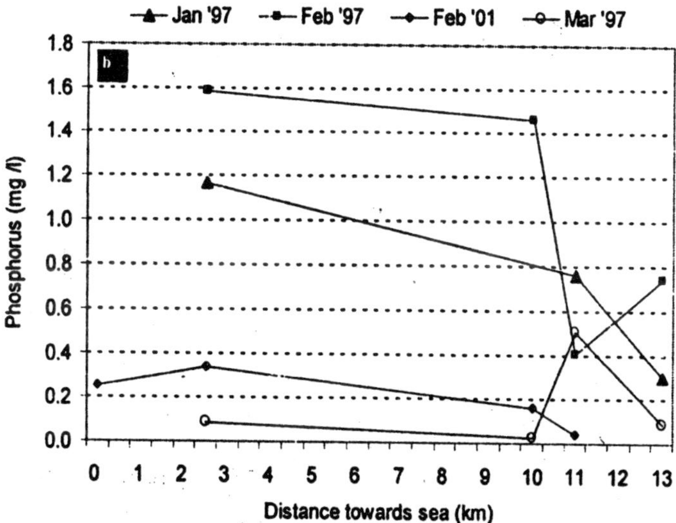

Fig. 6.19: Total phosphorus levels of Kaliveli lake during (a) monsoon, and (b) post monsoon

Plate 6.1: The agriculture practice at Kaliveli includes allowing cattle to graze on land so that their dropping fertilize the soil. Please also take note of manure heaps lined up by farmers as additional fertilizer. The cumulative effect of such intensive use of organics rich in N and P would aggravate the europhication of the already 'obese' Kaliveli.

Plate 6.2: The organic pollutants contribute by the surrounding agriculture fields washed into the lake during the monsoon as runoff.

Conclusion

Kaliveli is a very shallow and seasonal wetland; the fresh water, part of the lake holds water for 3-4 months only. The water quality of the lake changes as rapidly as the quantity of the water over this period.

The chief factors that control the dynamics of the lake are (i) direct precipitation (ii) non-point source pollutants—both agriculture and storm run-off (iii) evaporation and (iv) proportionate admixing of the sea water through the estuary.

The Kaliveli watershed being rich in iron, contributes to the lake through runoff. The lake, based on the criteria of total phosphorus had been classified as hyper-eutrophic.

REFERENCES

APHA, 1998 *Standard Methods for the Examination of Water and Waste Water*, Washington DC 2005.

Banerjee A.C., and Roy N.C., 1966. *Observations on Some Physico-chemical Features of the Chilka Lake*, Indian J. Fish, 13:395-429.

BIS, 1994. *Indian Standards Quality Tolerences for Fresh Water for Fish Culture*, Bureau of Indian Standards, Manak Bhavan, New Delhi.

Chandran R., and Ramamoorthi K., 1984. *Hydrobiological Studies in the Gradient Zone of the Vellar Estuary*, Mahasagar 20:1-13.

Hutchinson G.E., 1975. *A Treatise on Limnology: Chemistry of Lakes*. John Wiley and Sons Inc.

Nurnberg K.G, 1996. *Trophic State of Clear and Coloured, Soft and Hard Water Lakes with Special Consideration of Nutrients , Anoxia, Phytoplankton and Fish*, Journal of Lake and Reservoir Management, Vol 12, Iss 4, pp. 432-447.

Sarwar S. G and Rifet, 1991. *Abiotic Functions of a Fresh Water Lotic Ecosystem of Kashmir (Dhoodh Ganga River)*. Poll.res.,Vol:10 (2), pp. 69-73.

Upadhyay S., 1988. *Physico-chemical Characteristics of the Mahanadi Estuarine System*, East Coast of India, Indian J. Mar.Sci. 17:19-23.

ANNEXURE

Table—6.1: Abbreviations used for the various parameters in the subsequent tables and their units of measurement

	Abbreviations	*Units*
pH	pH	pH units
EC	Electrical conductivity	mmhos
Alk	Alkalinity	mgl^{-1}
Aci	Acidity	mgl^{-1}
CaH	Calcium hardness	mgl^{-1}
TotH	Total hardness	mgl^{-1}
Cl	Chlorides	mgl^{-1}
NO_2	Nitrites	mgl^{-1}
NO_3	Nitrates	mgl^{-1}
NH_3	Ammonical nitrogen	mgl^{-1}
OrgN	Organic nitrogen	mgl^{-1}
$RePO_4$	Reactive phosphorus	mgl^{-1}
$TotPO_4$	Total phosphorus	mgl^{-1}
SO_4	Sulfates	mgl^{-1}
Fe	Iron	mgl^{-1}

Table—6.2: The location of sampling stations and their distance towards the sea

Location	*Sample tag*	*Distance towards the sea*
Kunimedu	SW1	0.0 Km
Anumandaikuppam	SW2	2.5 Km
Vandipalayam	SW3	5.0 Km
Vadagaram	SW4	10.0 Km
Mudaliarpet	SW5	11.0 Km
Erapanni	SW6	13.0 Km

Table—6.3: Surface water quality of Kaliveli wetland, December 1998

Sam No.	Place	PH	EC	Alk	Aci	CaH	TotH	Cl	NO_2	NO_3	NH_3	OrgN	$RePO_4$	$TotPO_4$	SO_4	Fe
SW_2	Anumandai	7.2	0.360	34	4	14	52	36	1.20	0.03	0.17	0.92	0.097	0.227	2.0	2.19
SW_4	Vadagaram	6.7	0.083	18	6	10	24	11	0.93	0.02	0.03	0.10	0.108	1.427	1.4	1.50
SW_5	Mudaliarpet	7.2	0.098	36	4	24	36	6	2.25	0.05	0.04	0.49	0.184	1.190	1.9	2.35
SW_6	Erapanni	7.4	0.167	48	6	22	42	21	1.25	0.02	0.03	0.35	0.112	1.108	1.4	1.25

Table—6.4: Surface water quality of Kaliveli wetland, January 1999

Sam No	Place	pH	EC	Alk	Aci	CaH	TotH	Cl	NO_2	NO_3	NH_3	OrgN	SO_4	Fe
SW1	Kunimedu	8.3	1.63	64	0	30	128	281	0.0023	0.108	0.48	0.112	3.7	0.47
SW2	Anumandai	8.9	1.83	64	0	38	164	366	0.0047	0.095	–	–	4.177	0.64
SW4	Vadagaram	7.8	1.7	84	0	40	156	349	0.031	0.401	0.114	0.331	2.411	3.21
SW5	Mudaliarpet	8.3	2.25	92	0	52	196	459	0.014	0.226	0.39	0.257	3.264	1.64
SW6	Erapanni	8.2	2.72	128	0	78	264	408	0.0254	0.395	0.162	0.197	2.448	3.09

Table—6.5: Surface water quality of Kaliveli wetland, February, 1999

Sam No	*Place*	*pH*	*EC*	*Alk*	*Aci*	*CaH*	*TotH*	*Cl*	NO_2	NO_3	NH_3	*OrgN*	SO_4	*Fe*
SW1	Kunimedu	8.6	6.8	76	0	136	580	1589	0.0025	0.428	0.071	0.611	8.79	0.543
SW2	Anumandai	8.3	9.6	84	0	164	950	2345	0.002	1.744	0.212	6.7	8.023	2.199
SW4	Vadagaram	7.9	25.6	133	0	192	1430	6987	0.0015	1.652	0.044	0.757	9.418	1.633
SW5	Mudaliarpet	8.0	33	125	0	670	3660	11103	0.0006	2.509	0.108	0.693	9.348	1.007
SW6	Erapanni	8.0	36.6	129	0	690	4160	11534	0.0018	1.662	0.258	0.626	8.86	1.92

Table—6.6: Surface water quality of Kaliveli wetland, March 1999

Sam No	*Place*	*pH*	*EC*	*Alk*	*CaH*	*TotH*	*Cl*
SW1	Kunimedu	7.5	11	168	268	2000	2949
SW2	Anumandai	7.4	20	224	316	2640	4332
SW4	Vadagaram	7.7	40	256	348	3420	8480
SW5	Mudaliarpet	7.9	46	268	780	4240	11521
SW6	Erapanni	7.7	53	264	810	4540	13042

Table—6.7: Reconnaissance of surface water quality of Kaliveli wetland, March 2000

SamNo	*Place*	*pH*	*EC*
SW1	Kunimedu	8.1	4.90
SW2	Anumandai	8.9	5.00
SW3	Vandipalay	5.4	1.97
SW5	Mudaliarpet	7.9	3.23
PD1	Nadukuppam	6.3	0.05

Table—6.8: Reconnaissance of surface water quality of Kaliveli wetland, April 2000

SamNo	*Place*	*pH*	*EC*
SW1	Kunimedu	8	7.84
SW2	Anumandai	8.9	8.28
SW3	Vandipalay	8.1	11.29
SW5	Mudaliarpet	7.9	11.77
PD1	Nadukuppam	5.9	0.07

Table—6.9: Reconnaissance of surface water quality of Kaliveli wetland, June 2000

SamNo	*Place*	*pH*	*EC*
SW3	Vandipalay	7.8	36.2
SW5	Mudaliarpet	7.9	37.8
PD1	Nadukuppam	6.5	0.078

Table—6.10: Surface water quality of Kaliveli wetland, February 2001

SamNo	Place	PH	EC	Alk	Aci	CaH	TotH	Cl	NO2	NO3	NH3	OrgN	TotPO4	SO4	Fe
SW1	Kunimedu	7.6	12.85	20	80	19	2940	3445	3E-04	0.5724	0.274	0.2394	0.2511	278.03	0.6344
SW2	Anumandai	7.8	14.23	24	40	220	3274	3816	3E-04	0.3996	0.243	0.1938	0.3423	293.85	0.1888
SW3	Vandipalay	7.4	21.1	-	-	-	-	-	-	-	-	-	-	-	-
SW4	Vadagaram	7.6	12.62	20	60	200	2157	3323	9E-05	0.4212	0.467	0.3213	0.1582	279.8	0.407
SW5	Mudaliarpet	7.7	12.4	20	60	157	2196	3342	6E-05	0.108	0.535	0.357	0.0395	275.54	0.1686
PD1	Nadukuppam	6.2	0.06	-	-	-	-	-	-	-	-	-	-	-	-

Table—6.11: Surface water quality of Kaliveli wetland, April 2001

Sam No	Place	pH	EC	Alk	Aci	CaH	TotH	Cl	NO_2	NO_3	NH_3	$TotPO_4$	SO_4	Fe
SW3	Vandipalay	7.7	31.1	-	-	-	-	-	-	-	-	-	-	-
SW4	Vadagaram	8.2	24.2	-	-	-	-	-	-	-	-	-	-	-
SW5	Mudaliarpet	8.2	23.6	240	20	600	3140	10280	0.00019	0.34	0.8726	0.11	323.47	0.1345
PD1	Nadukuppam	6.2	0.13	-	-	-	-	-	-	-	-	-	-	-

Table—6.12: Surface water quality of Kaliveli wetland, June 2001

Sam No	*Place*	*PH*	*EC*	*Alk*	*Aci*	*CaH*	*TotH*	*Cl*	*NO_2*	*NO_3*	*NH_3*	*OrgN*	*$RePO_4$*	*$TotPO_4$*	*SO_4*	*Fe*
SW3	Vandipalay	7.7	39.8	200	16	1650	9800	29591	0.0027	0.98	0.07	1.392	0.0962	-	125.37	1.32
SW4	Vadagaram	7.8	41.0	100	20	1370	10000	26242	0.002	0.6	0.08	0.444	-	0.017	257.79	0.999
SW5	Mudaliarpet	8.0	39.7	180	56	1350	10000	5498	0.001	0.6	0.43	0.8889	0.0212	-	266.61	1.17
SW6	Erapanni	8.0	40.2	220	16	1800	11000	31740	0.0018	8.2	0.28	0.665	-	-	242.48	1.392
PD1	Nadukuppam	7.5	0.23	-	-	-	-	-	-	-	-	-	-	-	-	-
PD2	Ferdous Na	7.4	18.1	120	12	1100	6000	12996	0.0119	4.7	0.32	1.3426	-	0.309	31.86	7.33

Table—6.13: Surface water quality of Kaliveli wetland, August 2001

Sam No	*Place*	*pH*	*EC*	*Alk*	*Aci*	*CaH*	*TotH*	*Cl*	*NO_2*	*NO_3*	*NH_3*	*OrgN*	*$TotPO_4$*	*SO_4*
SW1	Kunimedu	7.5	7.72	21	11	660	2700	5849	0.0005134	10.0296	0.0899	0.6247	0.0081	53.26
SW3	Vandipalay	7.1	19.8	28	30	1100	9000	14180	0.0003757	38.446	0.3948	1.3701	0.0109	119.78
SW4	Vadagaram	7.3	19.7	-	-	-	-	-	-	-	-	-	-	-
SW5	Mudaliarpet	7.5	22.0	26	16	160	210	212	0.0000782	3.9999	0.0083	0.837	0.00566	526.74
SW6	Erapanni	7.7	21.8	-	-	-	-	-	-	-	-	-	-	-
PD1	Nadukuppam	6.8	0.43	-	-	-	-	-	-	-	-	-	-	-

Table—6.14: Surface water quality of Kaliveli wetland, October 2001

Sam No	Place	pH	EC	Alk	Aci	CaH	TotH	Cl	NO_2	NO_3	NH_3	OrgN	$RePO_4$	$TotPO_4$	SO_4	Fe
SW1	Kunimedu	7.2	1.12	-	24	224	172	296	0.0004403	2	0.1893	0.649	0.08	0.2	239	0.736
SW2	Anumandai	9.7	4.59	38	138	128	352	1626	0.0000374	1.408	0.1335	0.5246	0.056	0.186	246	3.936
SW3	Vandipalay	7.4	2.51	89	210	168	224	700	0.0002499	0.272	0.02	0.3787	0.0108	0.1	136	0.3776
SW4	Vadagaram	7.4	2.16	36	41	200	332	641	0.000406	1.808	0.1215	0.2274	0.072	0.18	225	0.848
SW5	Mudaliarpet	7.4	3.96	-	22	128	112	1153	0.0000374	0.656	0.0493	0.7599	0.02	0.16	201	0.3616
SW6	Erapanni	7.4	4.44	6	43	80	130	1429	0.0000748	0.816	-	0.535	0.0326	0.1	106	0.5184
PD1	Nadukuppam	6.1	0.16	-	-	-	-	-	-	-	-	-	-	-	-	-

7

Ground Water Quality of Kaliveli Lake And Its Surroundings

Abstract

In this chapter, a study on the ground water quality of Kaliveli and its surroundings has been presented. A total of 41 sampling stations, distributed around the lake, were monitored periodically during the year 2000 and 2001. pH, EC and other several essential parameters were analyzed and the results were discussed with respect to the drinking and irrigation water quality criteria of BIS. The areas with poor ground water quality had been identified using the spatial modelling of GIS.

Introduction

All through the human history, ground water has catered to the ever-increasing demands of domestic, industrial and irrigation sectors world-wide. Its role in providing water for drinking and irrigation in the regions such as the southern peninsular India has been particularly crucial due to the absence of perennial rivers in these regions.

The quality of ground water in an area is essentially a function of initial composition of water, precipitation, land use and the natural geology of the area. Activities, natural and anthropogenic, affect the regional ground water quantity and quality to a great extent. In this context, we have conducted an assessment of the ground water quality of Kaliveli and its surroundings.

Materials and Methods

Materials

The materials used in the present study are the same as presented in the chapter 6.

Methods

Ground Water Quality Monitoring

A reconnaissance of ground water quality monitoring was done in the year 2000 during the months of March, April and June. Subsequently an extensive ground water quality monitoring was done on alternate months between February 2001 and October 2001, so as to check the seasonal variations of the ground water.

A total of 41 ground water sampling stations were established for the present study (Table 7.1 and Figure 7.1). These included mostly the domestic hand-pumped bore wells and a few motorized bore wells, which are used for irrigation. These sampling stations were chosen in such a way that they are spread evenly around the Kaliveli lake. For GIS applications, these sites were identified precisely using the Survey of India (SOI) toposheets.

Water Quality Parameters

The analytical procedures adopted for the analyses of water quality were as detailed in the standard methods for examination of water and waste water (APHA, 1998).

The details of the methods used for estimating the various water quality parameters were as detailed in the chapter 10.

The villagers surrounding the lake depend exclusively on the ground water for drinking and irrigation purposes. Thus, the water quality results of ground water quality were compared with the water quality criteria of BIS for drinking water (1991) and irrigation (1974).

Table—7.1: Details of the ground water quality sampling stations

S. No	*Sam No.*	*Place*
1	GW1	Ichchangadu
2	GW2	Anpakkam
3	GW3	Kodur
4	GW4	Vilvanattam
5	GW5	Kaluperumbakam
6	GW6	Puthupattu
7	GW7	Puthupattu
8	GW8	Puthupattu
9	GW9	Puthupattu
10	GW10	Puthupattu
11	GW11	Ferdous Nagar
12	GW12	Kunimedu
13	GW13	Seyyankuppam
14	GW14	Seyyankuppam
15	GW15	Anumandaikuppam
16	GW16	Anumandaikuppam
17	GW17	Kilpettai
18	GW18	Kilpettai
19	GW19	Kalikuppam
20	GW20	Kalikuppam
21	GW21	Vadagaram
22	GW22	Kandadu
23	GW23	Mudaliarpet
24	GW24	Mudaliarpet
25	GW25	Mudaliarpet
26	GW26	Pachapaithankollai
27	GW27	Pudupakkam
28	GW28	Tirukanur
29	GW29	Nadukuppam
30	GW30	Vandipalayam
31	GW31	Vandipalayam
32	GW32	Vandipalayam
33	GW33	Vandipalayam
34	GW34	Talaganikuppam
35	GW35	Talaganikuppam
36	GW36	Talaganikuppam
37	GW37	Devanandal
38	GW38	Katterimedu
39	GW39	Velur
40	GW40	Pudukuppam
41	GW41	Dhanamaniyam

Data Presentation

The water quality data has been presented as Annexure at the end of this chapter (Table 7.2 till Table 7.14).

Spatial Modelling

For the data input MapInfo Professional 5.5, and for spatial modeling, grid generation and contour mapping Vertical Mapper 2.6 were used.

Results and Discussion

pH

During the monsoon pH values varied as low as 3.5 at Ferdous Nagar (GW11) during the month of Oct' 01, and 8.0 at Katterimedu (GW38) during June '01, and Kalikuppam (GW20) during Aug '01 (Figs. 7.7, 7.8 and 7.9).

An average of 38% of the samples were below the permissible limits of water quality criteria of BIS for drinking water (1991), and rest of the samples were within the permissible limits. Similarly, as per the irrigation water quality criteria of specified by BIS (1974), 92% of the samples were within the permissible limits and the remaining 8% were below the permissible limits.

During the post monsoon period, pH of ground water varied between 4.4 at Ferdous Nagar (GW11) during April '00, and 8.0 at Anumandaikuppam (GW16) during Apl '00 (Fig. 7.3).

As per the water quality criteria of BIS for drinking water (1991), an average of 60% of the samples were below the permissible limits and the rest, 40% were within the permissible limits. There was a marked change in the pH of ground water during the monsoon. Nearly, 22% of the sampling sites were highly vulnerable to the seasonal fluctuations.

The geospatial modelling of ground water pH at Kaliveli (Fig. 7.10) reveals that precipitation, and the subsequent leaching of the soil constituents, and evaporation are the primary factors that determine the ground water quality.

The severe to mild acidic condition of the ground water, found at some of the sites, can lead to several consequences such as leaching of metals, especially iron. The local people also complained of sharp taste of the ground water.

As per the water quality criterion of BIS for irrigation water (1974), nearly 78% of the samples were within the permissible limits and the rest of them fall below the permissible limits.

Alkalinity

The lowest alkalinity in the ground water observed during the monsoon was 2 mg l^{-1}, at Kilpettai (GW17) during Oct '01 and the maximum was 480 mg l^{-1} at Kaluperumbakam (GW5) during the month of Jun '01 (Figs. 7.13, 7.15).

While, during the period of post monsoon the lowest levels of alkalinity, 8 mg l^{-1}, was observed at at Seyyankuppam (GW13) during Feb '01, and a maximum of 100 mg l^{-1} at Vadagaram (GW 21) during the month of Apl '01 (Figs. 7.11, 7.12).

As per the water quality criteria of BIS for drinking water (1991) all the samples were below the permissible limits both during the monsoon and post monsoon.

Acidity

During monsoon period, acidity levels in the ground water varied between 12 mg l^{-1} (GW23, GW36, Jun '01; GW33, Aug '01) and 220 mg l^{-1} (GW3, Oct '01) (Figs. 7.18, 7.20).

The acidity levels during the post monsoon ranged between 20 mg l^{-1} (GW6, GW13, GW23, Apl '01) and 200 mg l^{-1} (GW17, Feb '01) (Figs. 7.16, 7.17).

The general trend observed was that, the acidity of ground water tends to increase gradually as the monsoons pick up; and again as the post monsoon begins the ground water acidity decreases. This could be attributed to the following reasons: (i) the recharging of the ground water aquifers during the monsoons result in dilution of the alkaline salts, (ii) the percolation of rain water might bring in the acidic constituents of soils such as humic acids and iron and (iii) the lake, which is the largest source for ground water recharge, becomes mildly acidic during the monsoons.

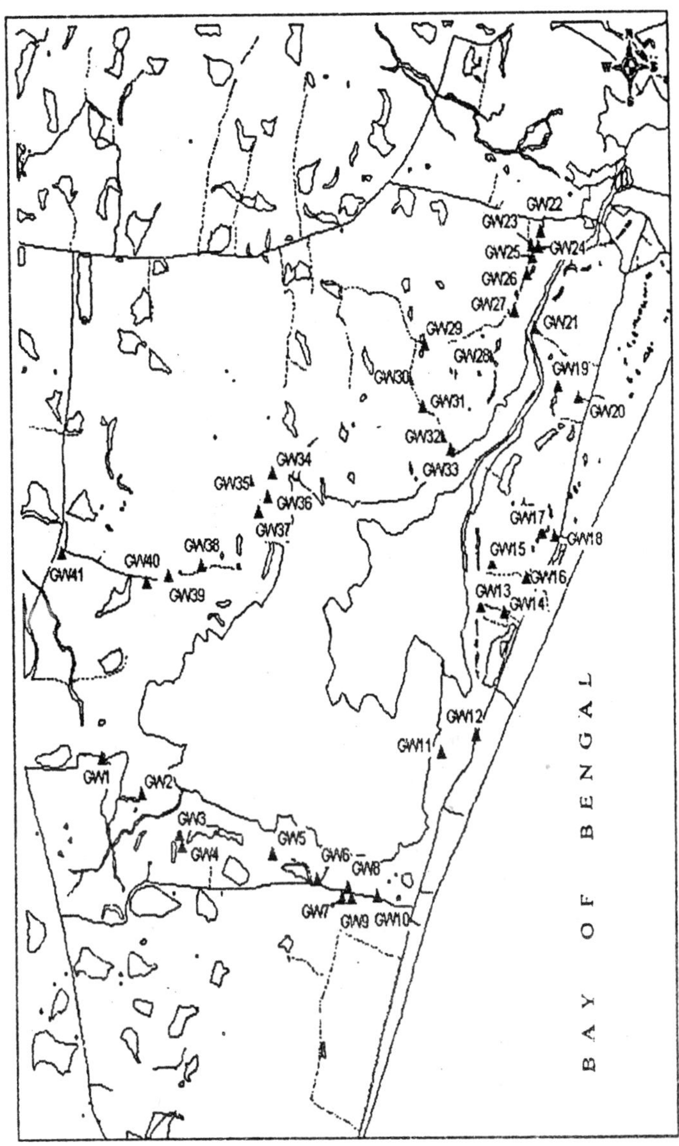

Fig. 7.1: Ground water sampling stations in and around Kaliveli

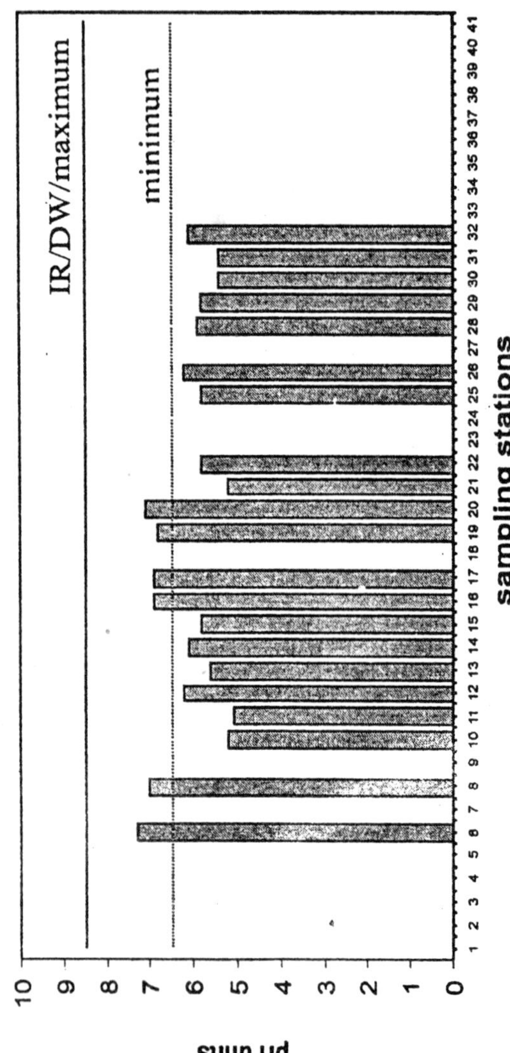

Fig. 7.2: pH of ground water, March 2000

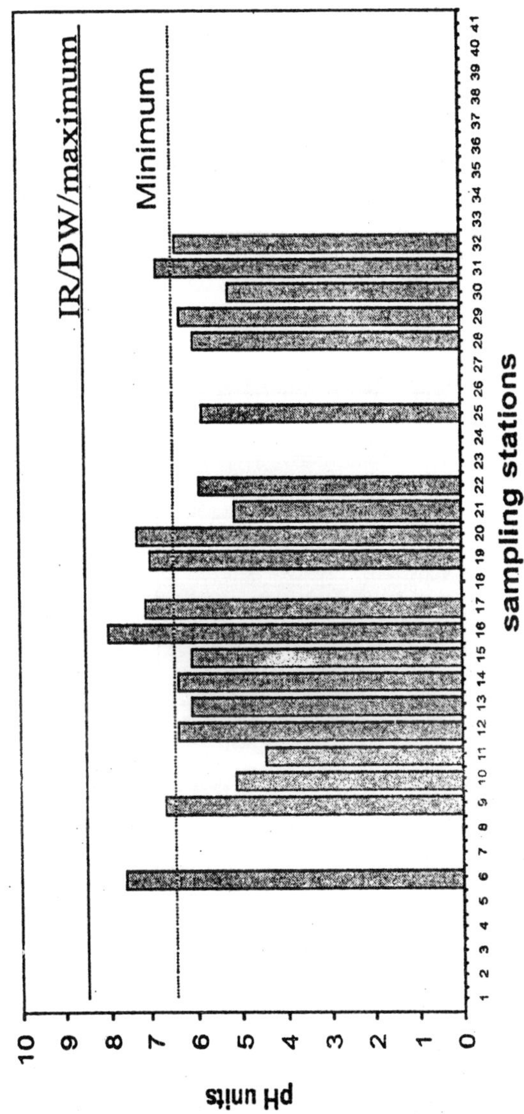

Fig. 7.3: pH of ground water April 2000

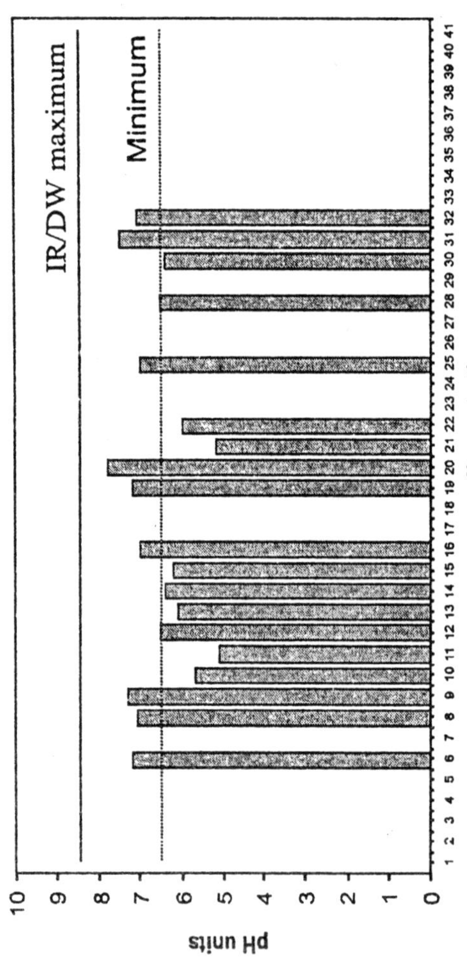

Fig. 7.4: pH of ground water, June 2000

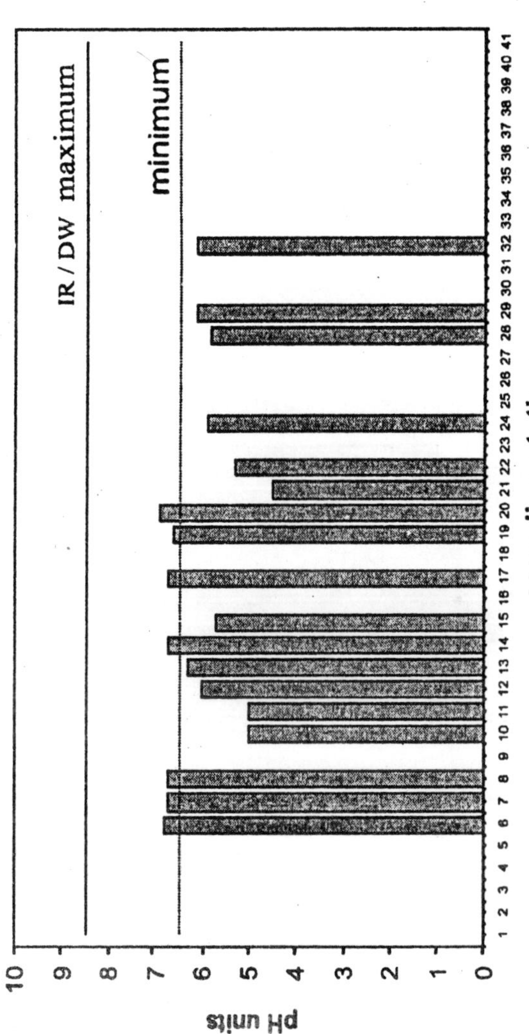

Fig. 7.5: pH of ground water, February 2001

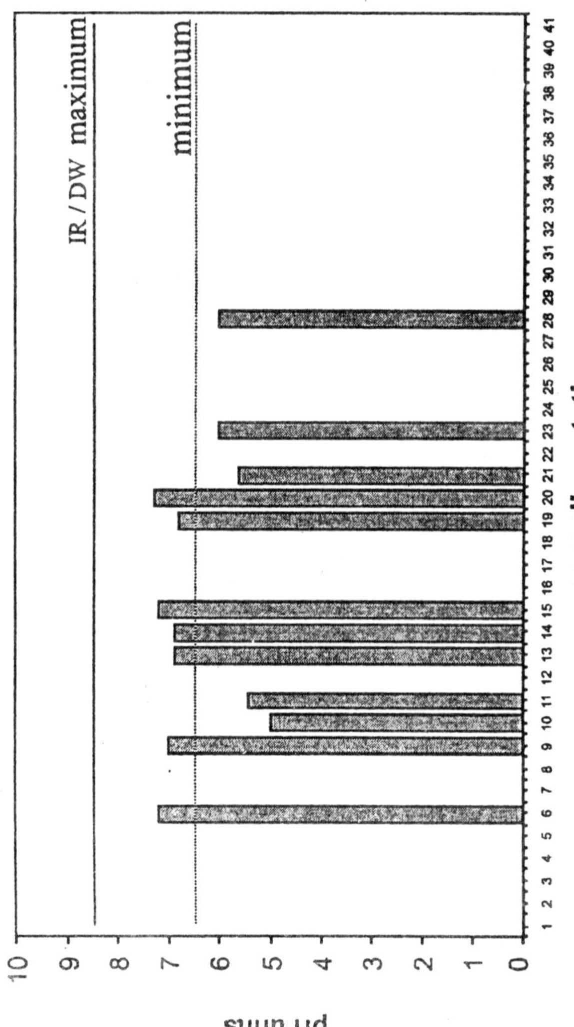

Fig. 7.6: pH of ground water, April 2001

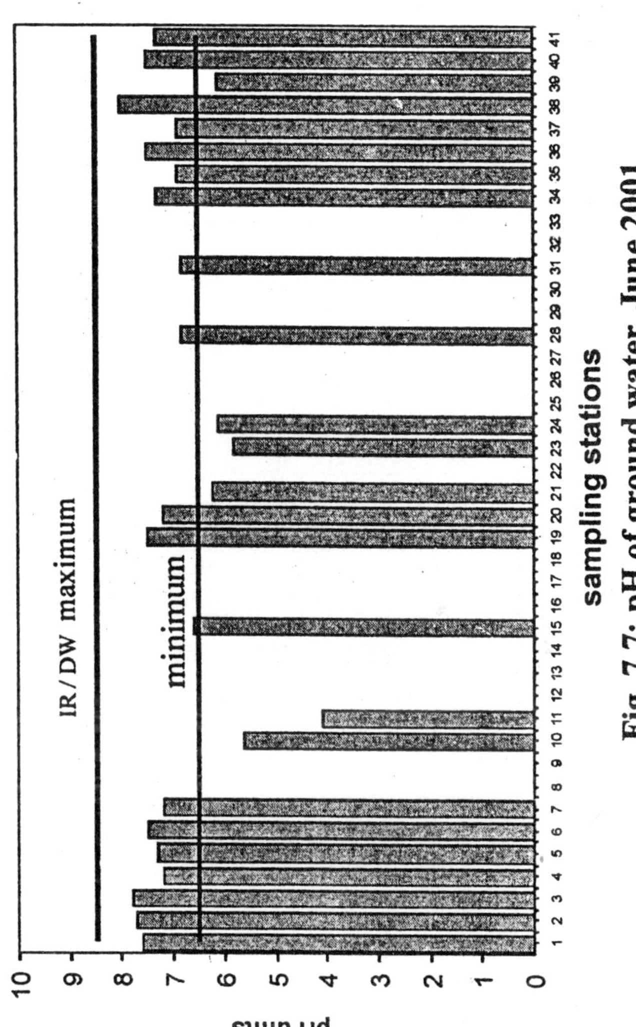

Fig. 7.7: pH of ground water, June 2001

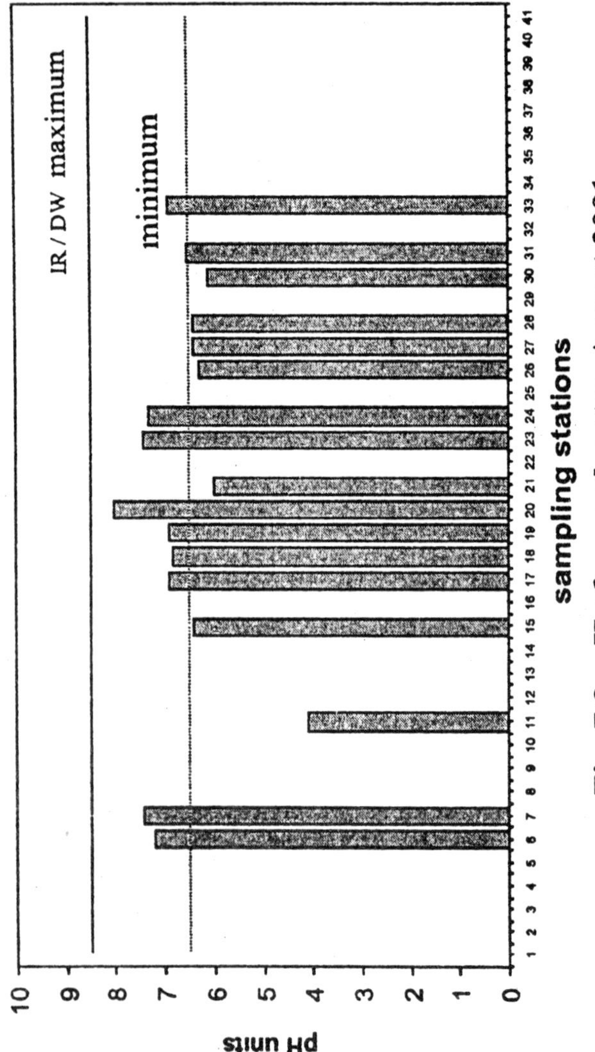

Fig. 7.8: pH of ground water, August 2001

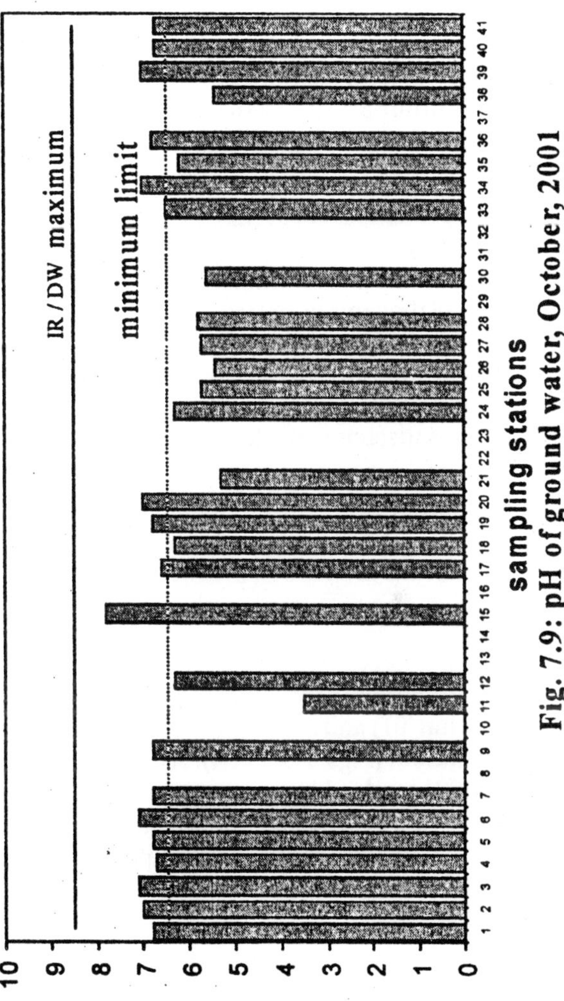

Fig. 7.9: pH of ground water, October, 2001

Electrical Conductivity

The EC of the ground water during the period of monsoon ranged as low as 0.12 mmhos at Talaganikuppam (GW36) during Jun '01, and as high as 3.74 mmhos at Anumandaikuppam (GW15) during Jun '00 (Figs. 7.23, 7.26, 7.28). The ground water of Kaliveli had been classified into four catergories based on Garg (1980). During monsoon, a majority of the sampling stations, nearly 51% fell in the category of C3, with medium salinity hazard. The sampling stations, GW6 (Puthupattu), GW15 (Anumandaikuppam), and GW21 (Vadagaram), pose a high salinity hazard and the water is unsatisfactory.

During the post monsoon period the EC varied between 0.12 mmhos at Vandipalayam (GW30) during Mar '00, and 3.72 mmhos at Anumandaikuppam (GW15) during Apl '01 (Figs. 7.21, 7.22, 7.24, 7.25). A majority of the sampling stations, an average of 60%, fell in the category of C3, with medium salinity hazard. The sampling stations, GW6 (Puthupattu), GW15 (Anumandaikuppam), GW19 (Kalikuppam), and GW21 (Vadagaram), pose a high salinity hazard and the water is unsatisfactory during the post monsoons.

The contours of EC (Fig. 7.29) indicate that monsoons contribute to the mild salinity of the aquifers in Kaliveli region due to the leaching of decomposed organic matter, humic substances and the salts of soils that dissolve during the infiltration of water.

Chlorides

The chloride of ground water was observed to vary between 15 mg l^{-1} (GW23, Jun '01) and 23042 mg l^{-1} (GW11, Aug '01) during the period of monsoon (Figs. 7.32, 7.34). As per the water quality criteria of BIS (1991) for drinking water only one sampling station, GW11 (Ferdous Nagar) was above the permissible limits; and as per the criteria of irrigation water two sampling stations, GW11 (Ferdous Nagar) and GW24 (Mudaliarpet) were above the permissible limits.

The chloride levels during post monsoon were as low as 46 mg l^{-1} (GW13, Feb '01) and as high as 23042 mg l^{-1} (GW17, Aug '01) (Figs. 7.30 and 7.33). As per the water quality criteria of BIS (1991) for drinking and irrigation water, only one sampling station, GW17 (Kilpettai) was above the permissible limits.

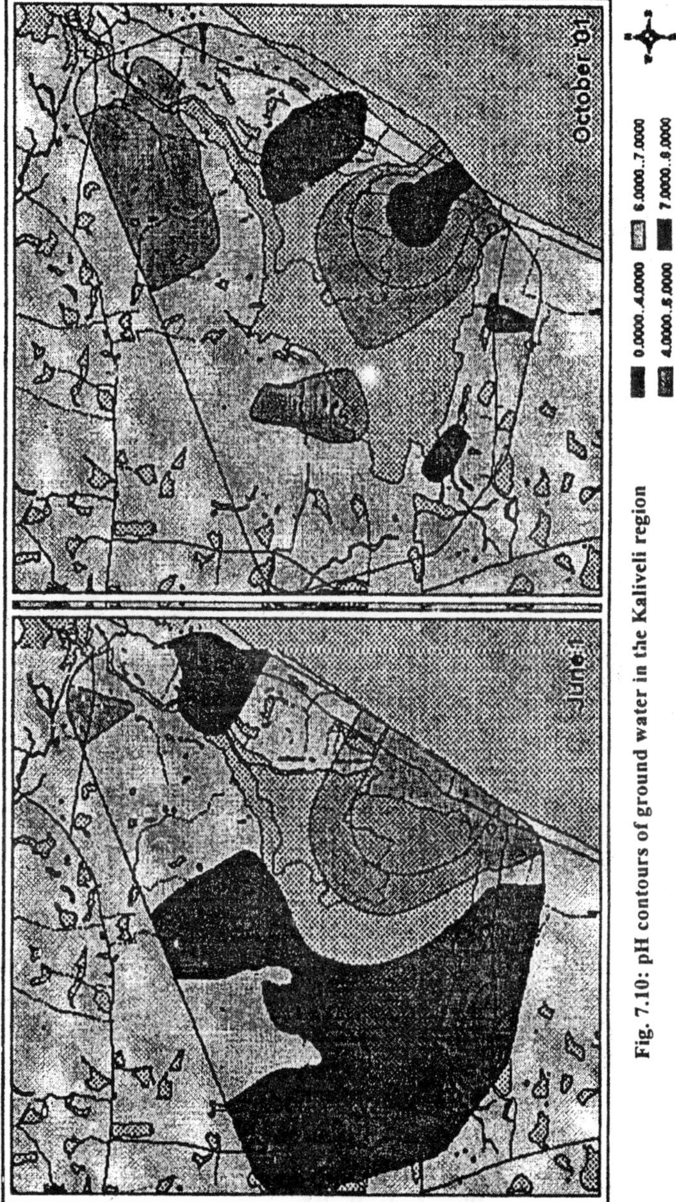

Fig. 7.10: pH contours of ground water in the Kaliveli region

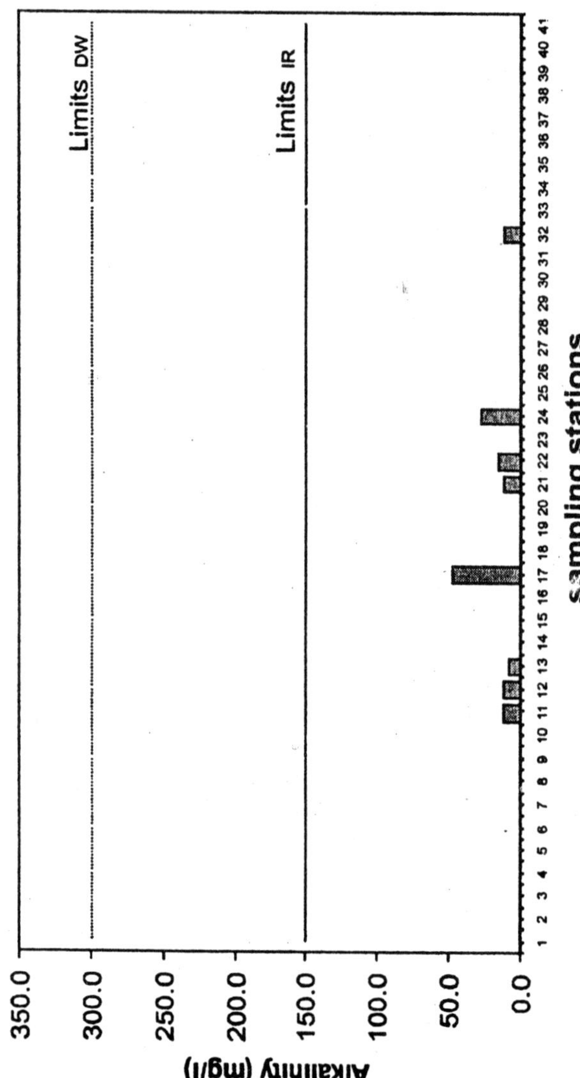

Fig. 7.11: Alkalinity of ground water, February 2001

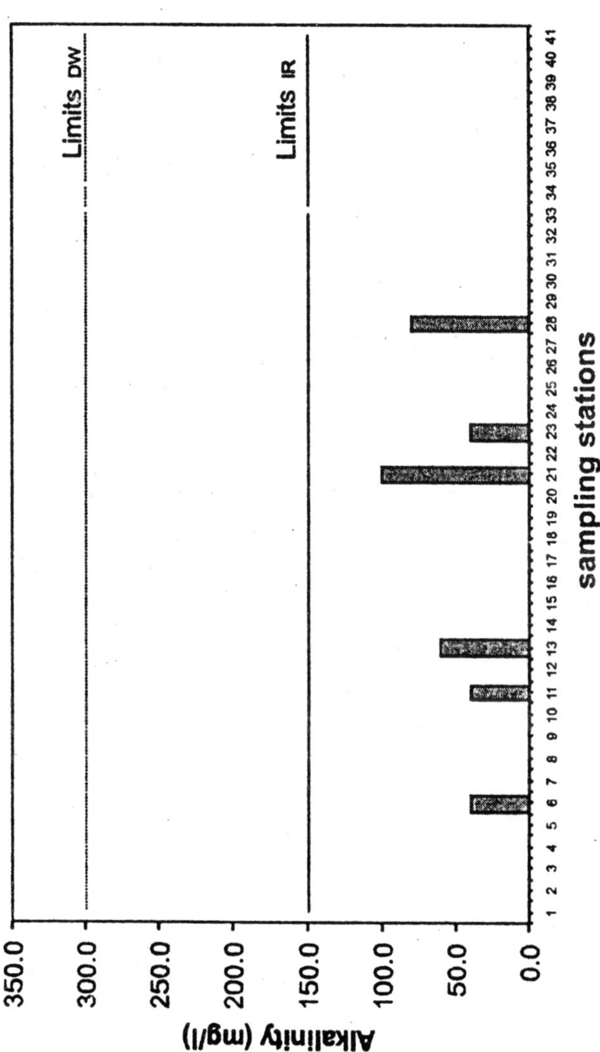

Fig. 7.12: Alkalinity of ground water, April 2001

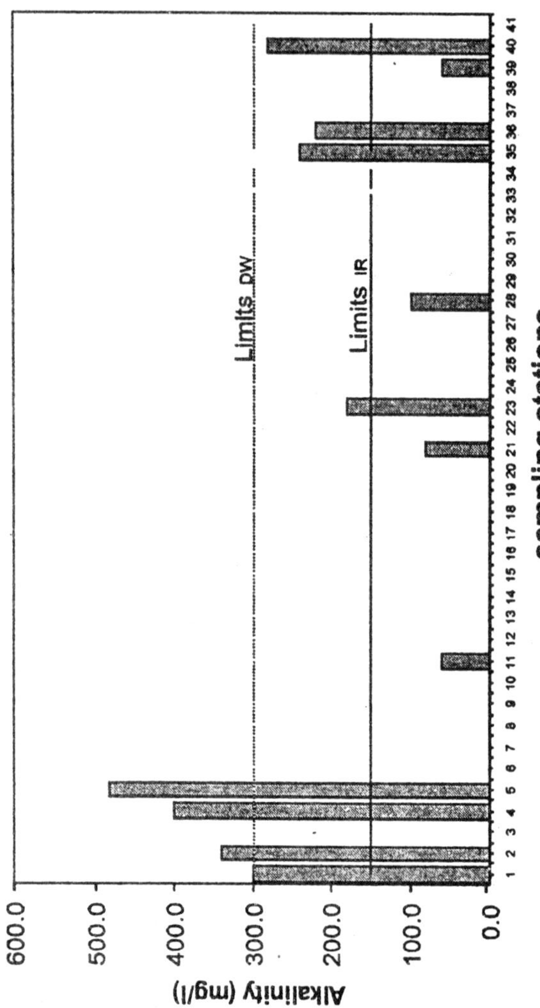

Fig. 7.13: Alkalinity of ground water, June 2001

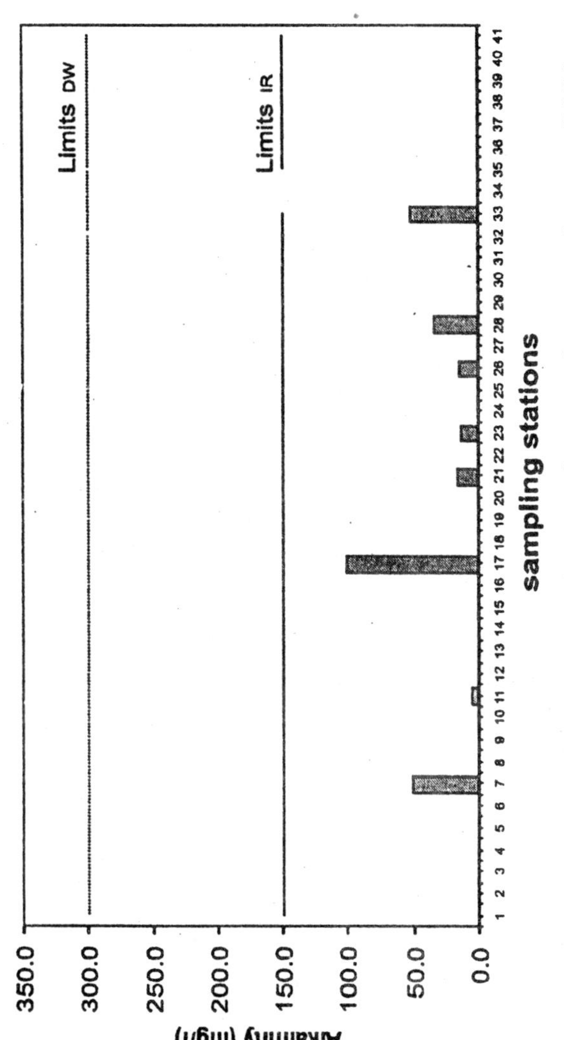

Fig. 7.14: Alkalinity of ground water, August 2001

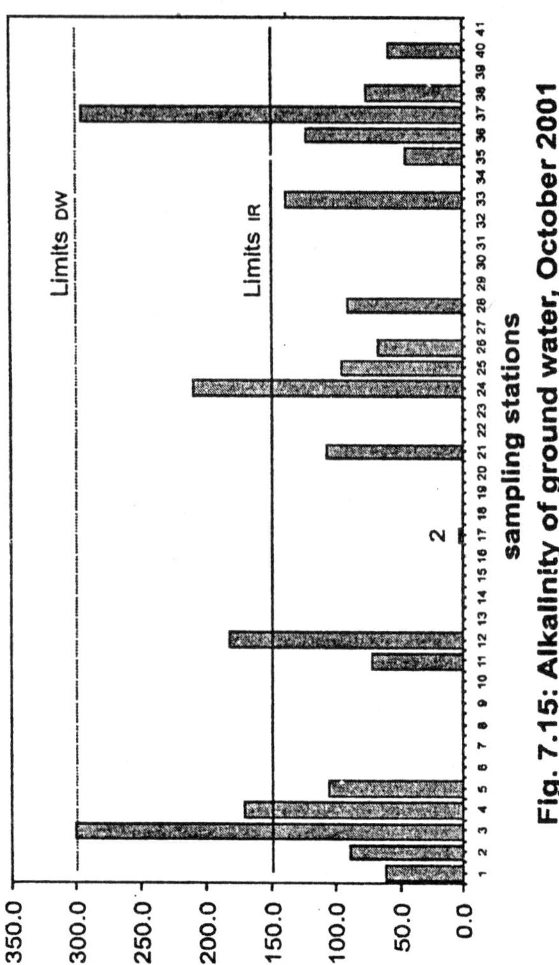

Fig. 7.15: Alkalinity of ground water, October 2001

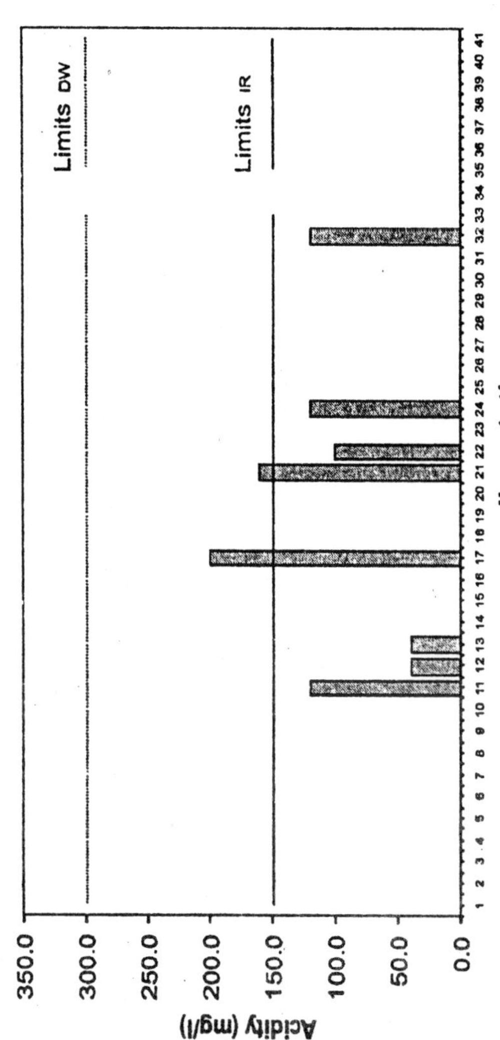

Fig.7.16: Acidity of ground water, February 2001

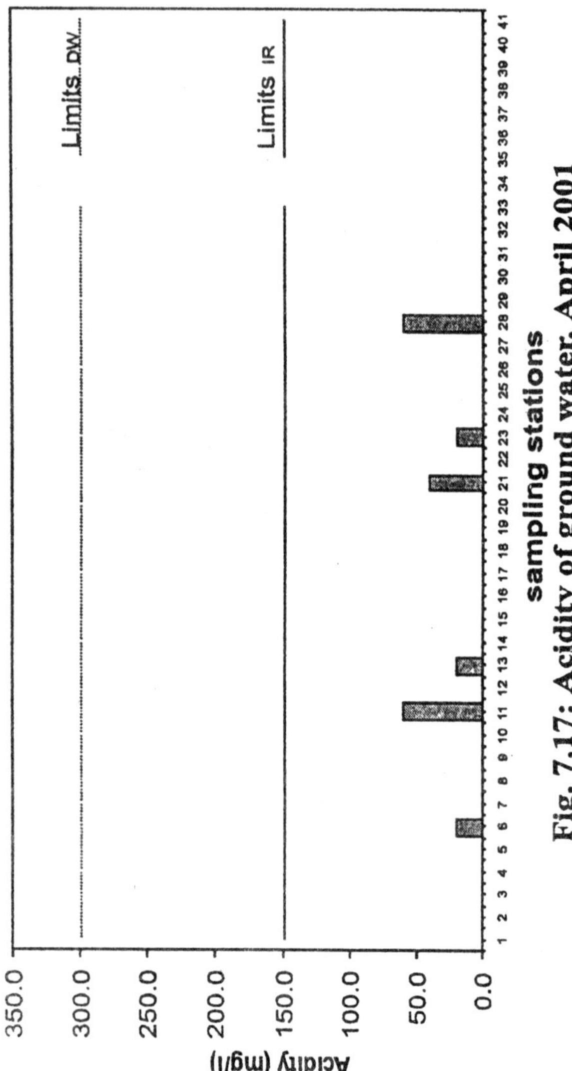

Fig. 7.17: Acidity of ground water, April 2001

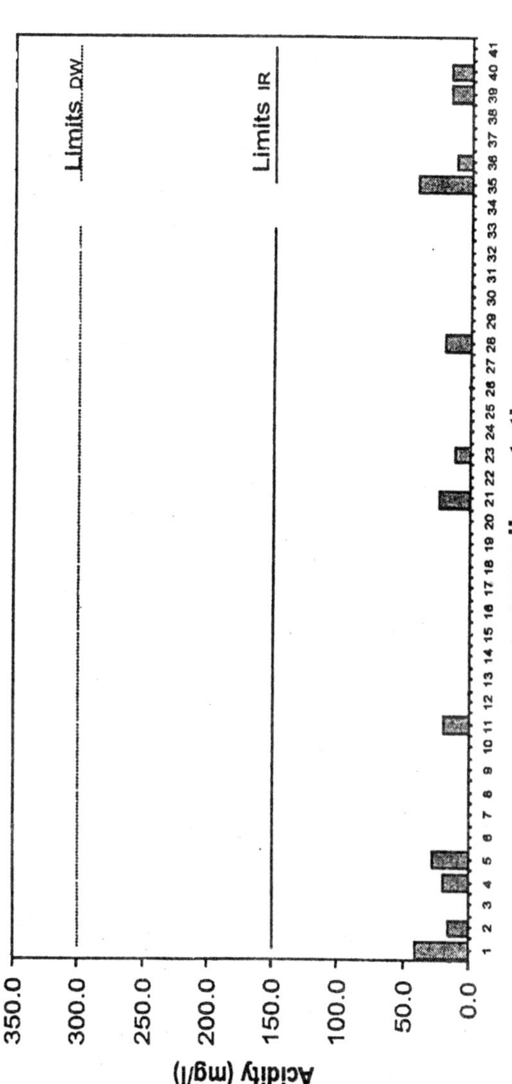

Fig. 7.18: Acidity of ground water, June 2001

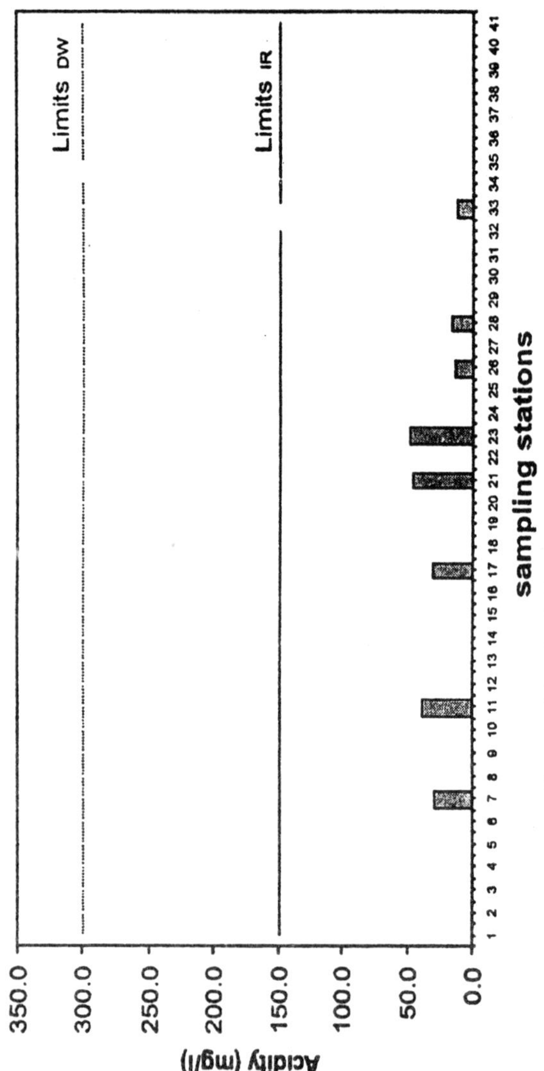

Fig. 7.19: Acidity of ground water, August 2001

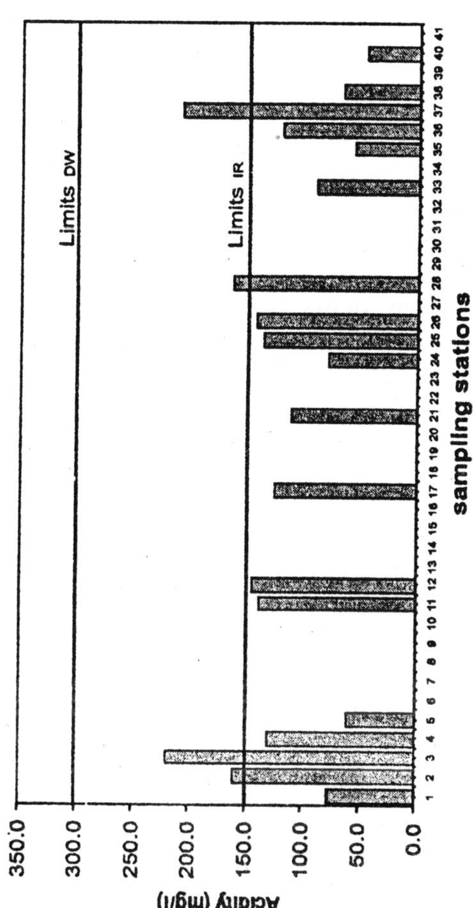

Fig. 7.20: Acidity of ground water, October 2001

Total Hardness

The total hardness ranged between 22 mg l^{-1} (GW21, Vadagaram, Aug '01) and 14500 mg l^{-1} (GW11, Ferdous Nagar, Aug '01) during the monsoon (Figs. 7.37-7.39). An average of 96% of the total sampling stations has total hardness values within the permissible limits of drinking water criteria of BIS (1991).

During post monsoon the total hardness ranged between 22 mg l^{-1} (GW22, Kandadu, Feb '01) and 5000 mg l^{-1} (GW21, Vadagaram, Feb '01) (Figs. 7.35, 7.36). Unlike monsoon period, fewer sampling stations, 81% of them met with the drinking water criteria of BIS (1991).

Calcium Hardness

During monsoon, all the samples were within the permissible of drinking water criteria of BIS (1991). The calcium hardness levels in the ground water varied between 2 mg l^{-1} at GW21 (Vadagaram, Aug'01) and 280 mg l^{-1} at GW4 (Vivanattam, Oct '01) (Figs. 7.42-7.44).

The lowest calcium hardness, observed during the post monsoon was 9 mg l^{-1} at Seyyankuppam (GW13, Feb '01), and the highest, 180 mg l^{-1} at Puthupattu (GW6, Apl '01) (Figs. 7.40, 7.41). During this period about 81% of the samples were below the permissible limits of BIS for drinking water quality criteria (1991). The calcium levels in the ground water were observed to be higher during the monsoon period than the post monsoon period.

The sampling stations with above the permissible limits of calcium hardness for the drinking water criteria were GW4 (Vilvanattam), GW5 (Kaluperumbakkam), GW7 (Puthupattu), GW 17 (Kilpettai), GW21 (Vadagaram) and GW40 (Pudukuppam).

Sulfates

The lowest level of sulfates, 0.48 mg l^{-1}, during monsoon was observed at Tirukanur (GW28, Aug '01) and the highest, 360 mg l^{-1}, at Kunimedu (GW12, Oct '01) (Figs. 7.47-7.49).

During post monsoon the sulfate levels in ground water varied between 7.8 mg l^{-1} (GW22, Kandadu, Feb '01) and 271 mg l^{-1} (GW11, Ferdous Nagar, Feb '01) (Figs. 7.45, 7.46).

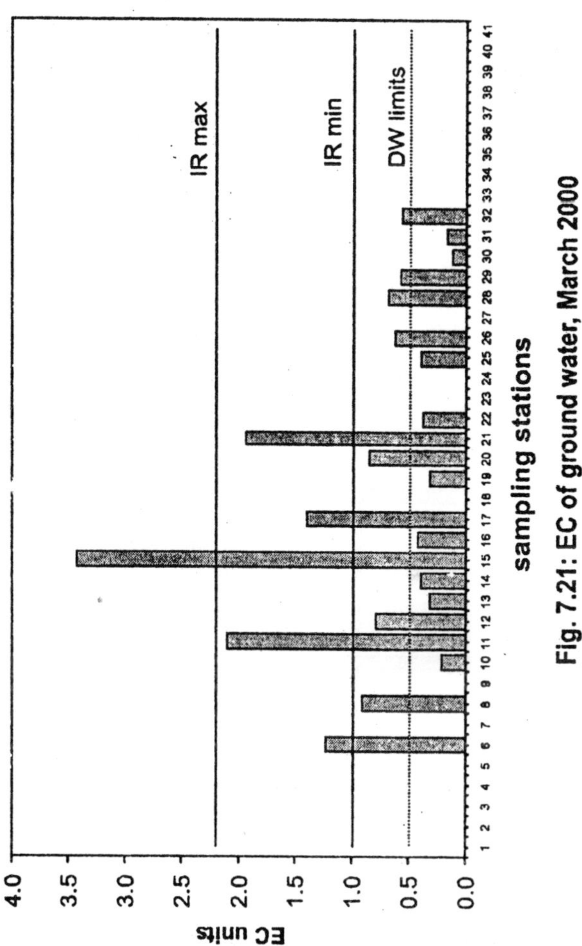

Fig. 7.21: EC of ground water, March 2000

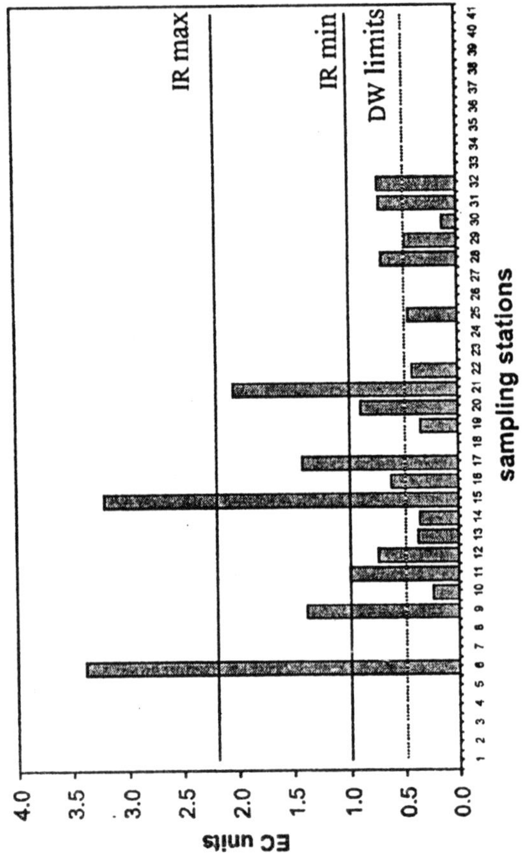

Fig. 7.22: EC of ground water, April 2000

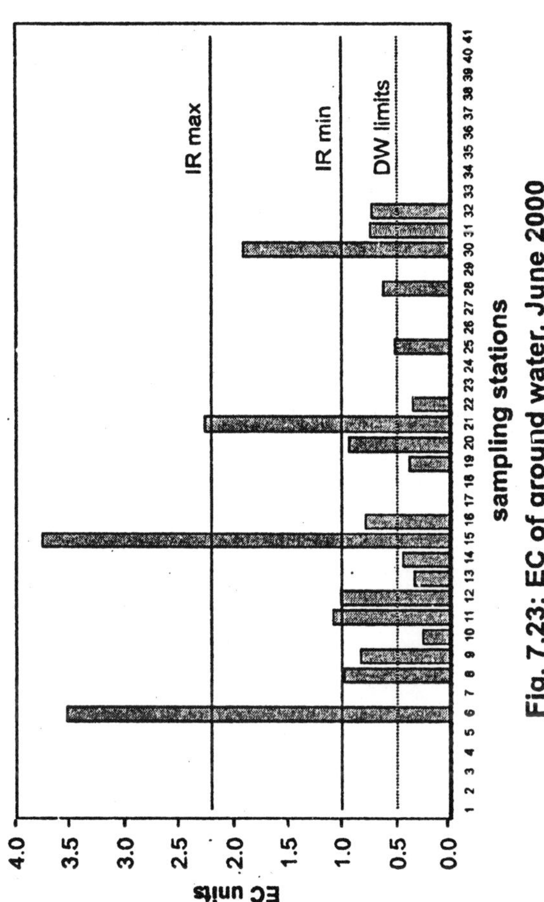

Fig. 7.23: EC of ground water, June 2000

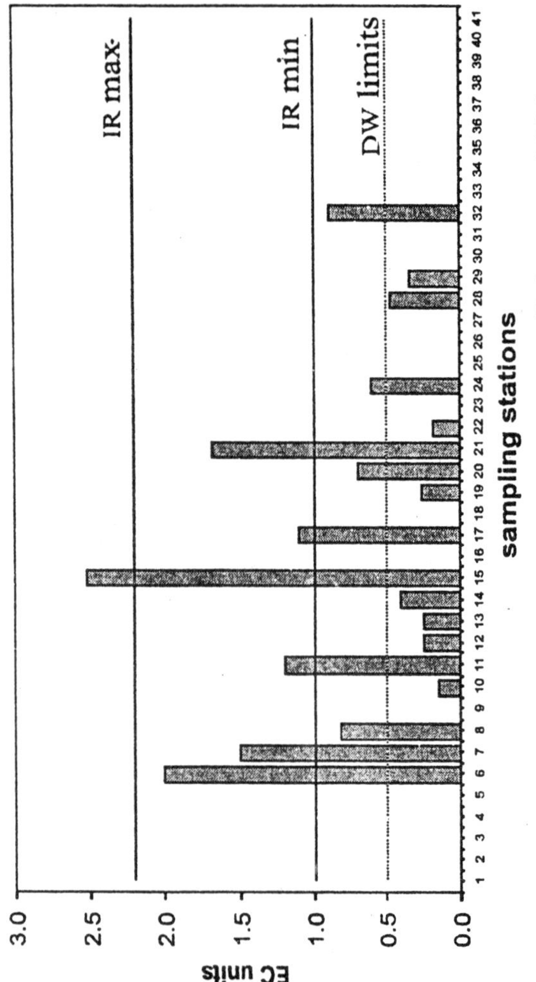

Fig. 7.24: EC of ground water, February 2001

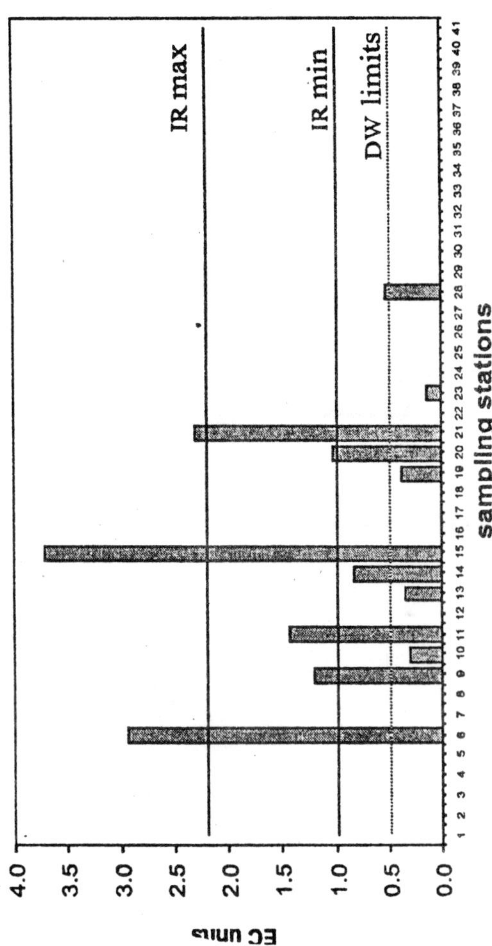

Fig. 7.25: EC of ground water, April 2001

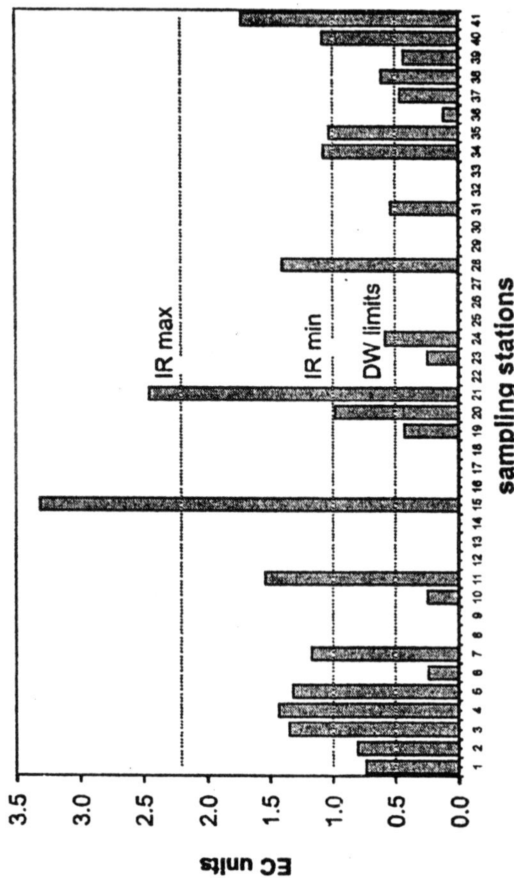

Fig. 7.26: EC of ground water, June 2001

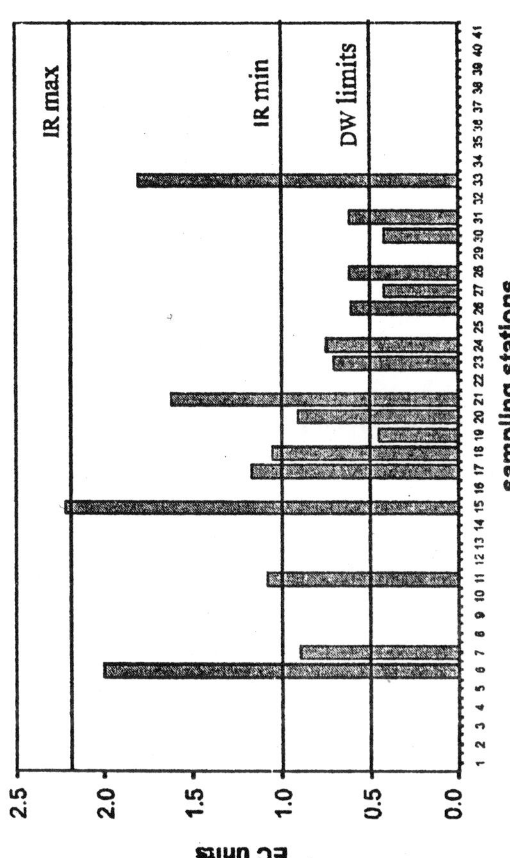

Fig. 7.27: EC of ground water, August 2001

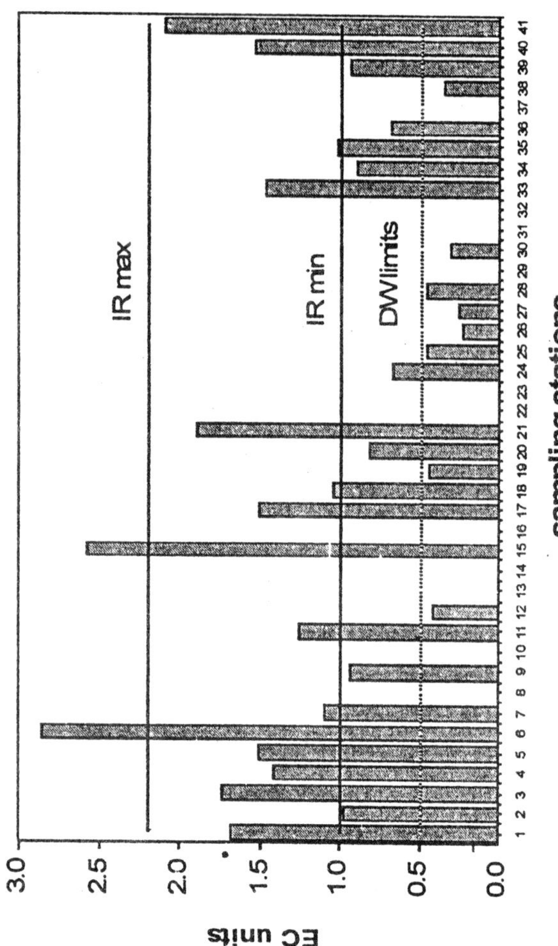

Fig. 7.28: EC of ground water, October 2001

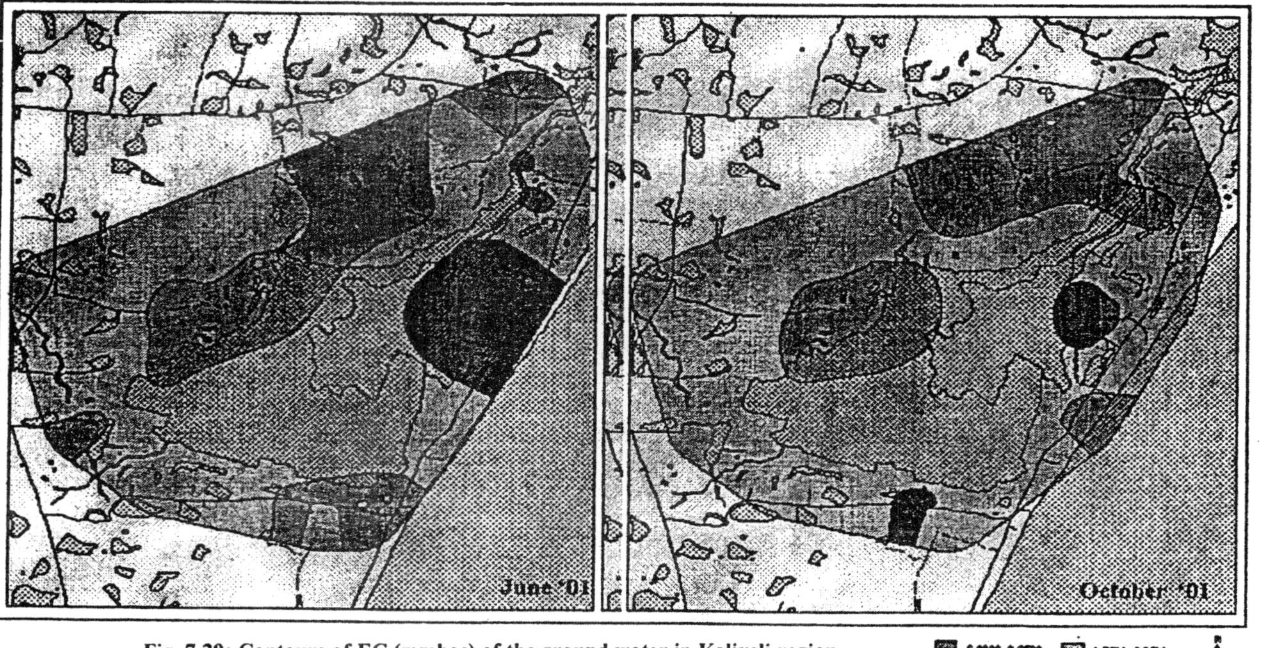

Fig. 7.29: Contours of EC (mmhos) of the ground water in Kaliveli region

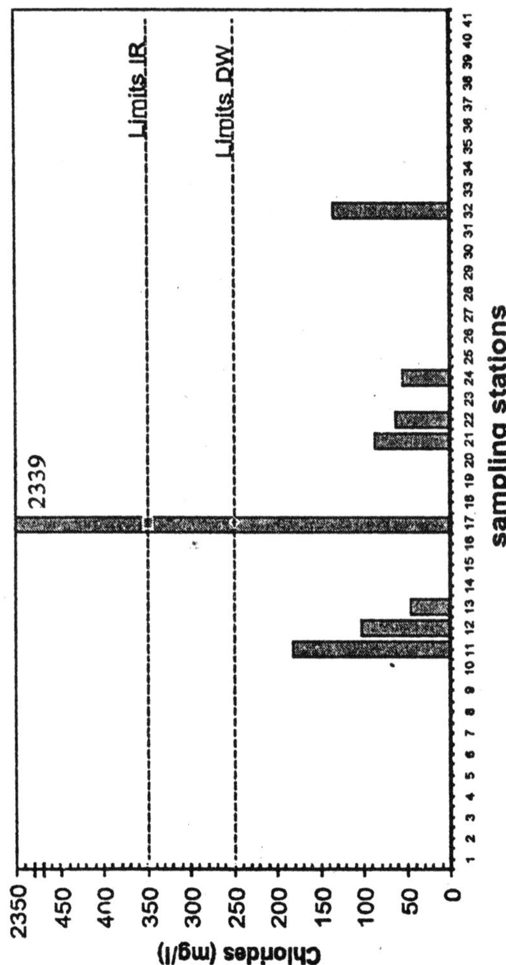

Fig. 7.30: Chloride of ground water, February 2000

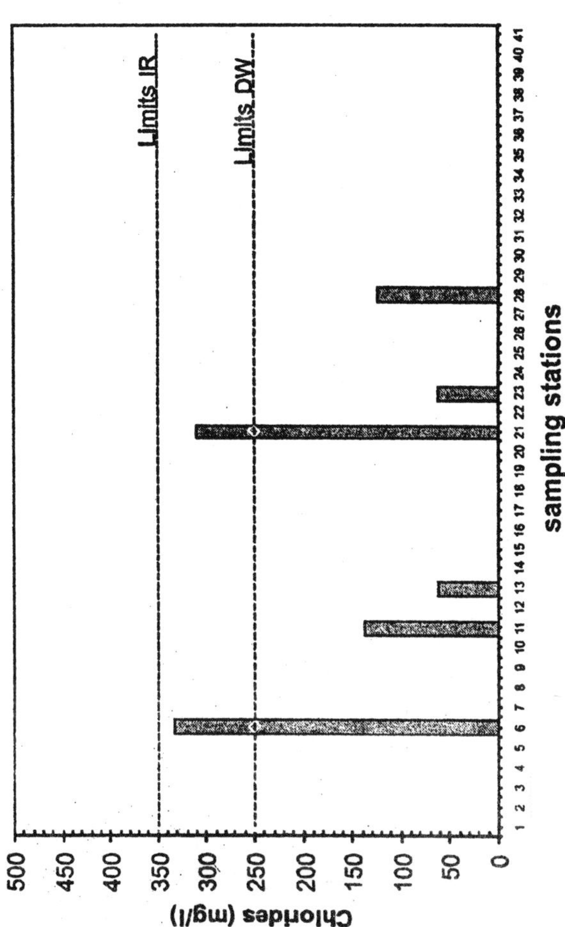

Fig. 7.31: Chloride of ground water, April 2000

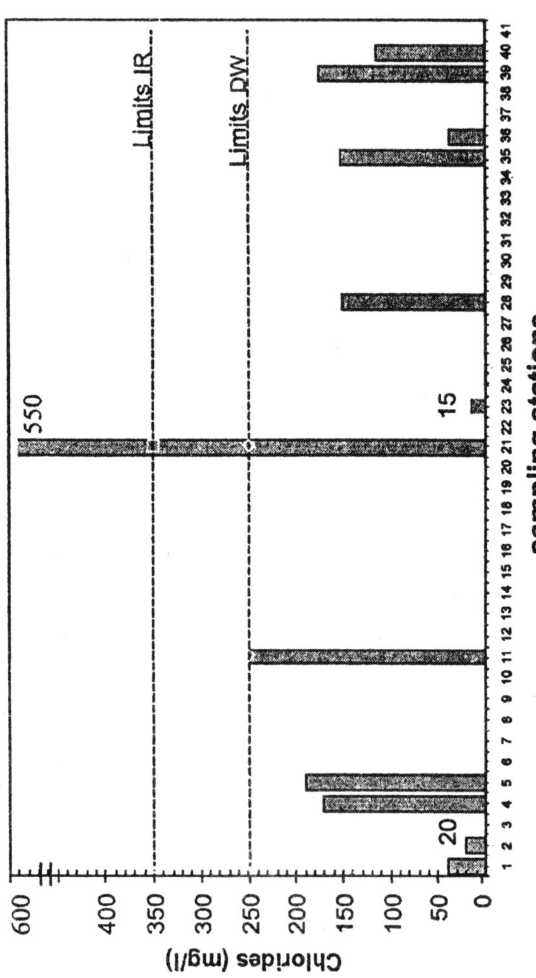

Fig. 7.32: Chloride of ground water, June 2001

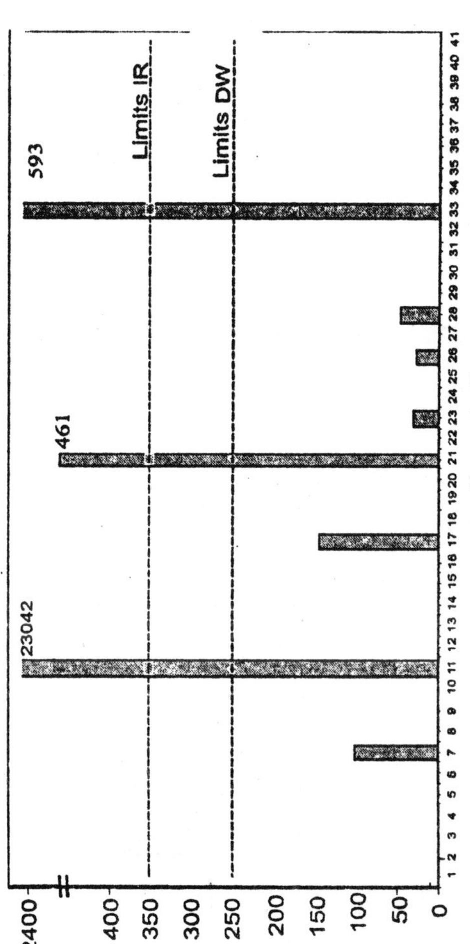

Fig. 7.33: Chloride of ground water, August 2001

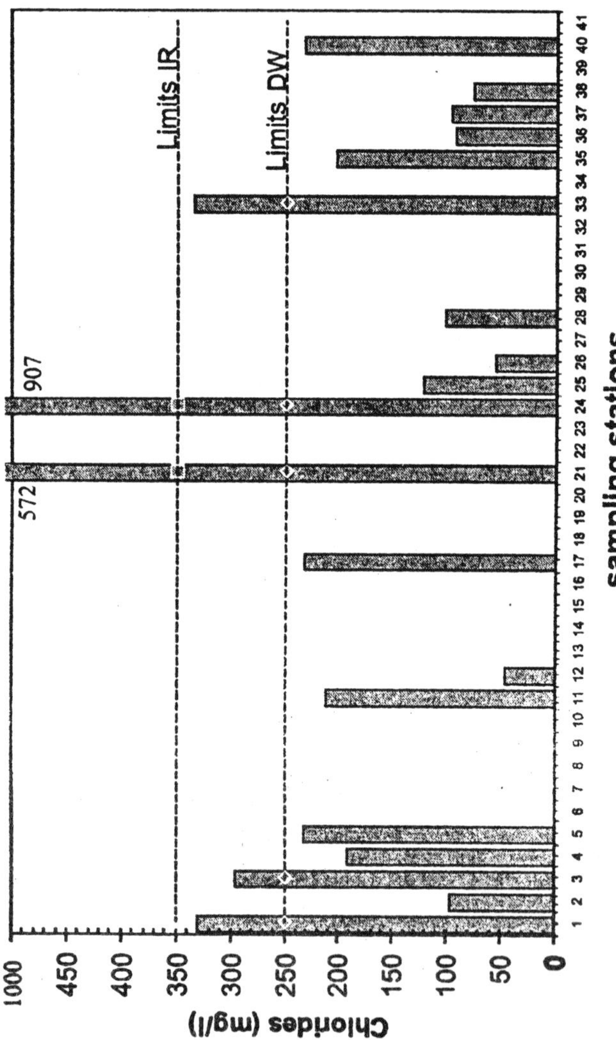

Fig. 7.34: Chloride of ground water, October 2001

All the samples were within the permissible limits of BIS for drinking (1991) and irrigation water (1974) criteria, both during the monsoon and post monsoon.

Total Iron

During monsoon, total iron in the ground water varied between 0.37 mg l^{-1} (GW35, Talaganikuppam, Oct '01) and 6.45 mg l^{-1} (GW28, Tirukanur, Jun '01) (Figs. 7.52, 7.53). During this period nearly 54% of the samples were above the permissible limits of drinking water criteria (BIS, 1991). The sampling stations that were above the permissible limits were GW2, GW21, GW23, GW28, GW35, GW36, during the month of Jun '01 and GW2, GW4, GW1, GW12, GW24, GW25, GW26, GW28, GW33, GW36, and GW37 during Oct '01.

The total iron levels, during the post monsoon, ranged between 0.0356 mg l^{-1} (GW24, Mudaliarpet, Feb '01) and 0.5752 mg l^{-1} (GW13, Seyyankuppam, Feb '01) (Figs. 7.50, 7.51) and all the samples were below the permissible limits of drinking water criteria (BIS, 1991).

The high iron contents in the ground water of Kaliveli can be due to the agricultural activities, low water level in the catchment, increased diffusion of iron from sediments of lower oxygen near sediments (Sarwar and Rifet, 1991).

Nitrogen

Nitrites: During monsoon, nitrites in the ground water ranged between 0.0000017 mg l^{-1} (GW11, Ferdous Nagar, Aug '01) and 0.4059 mg l^{-1} (GW36, Talaganikuppam, Oct '01) (Figs. 7.56-7.58).

Nitrites in the ground water during post monsoon varied between 0.000003 mg l^{-1} (GW22, Kandadu, Feb '01) and 0.4039 mg l^{-1} (GW24, Mudaliarpet, Feb '01) (Figs. 7.54, 7.55).

Nitrates: The nitrate levels of the ground water varied between 0.4 mg l^{-1} (GW35, Talaganikuppam, Jun '01) and 72.4 mg l^{-1} (GW21, Vadagaram, Aug '01) during the monsoon (Figs. 7.61-7.63). In post monsoon, the range was 0.1782 mg l^{-1} (GW17, Kilpettai, Feb '01) and 5.1894 mg l^{-1} (GW13, Seyyankuppam, Feb '01) (Figs. 7.59, 7.60).

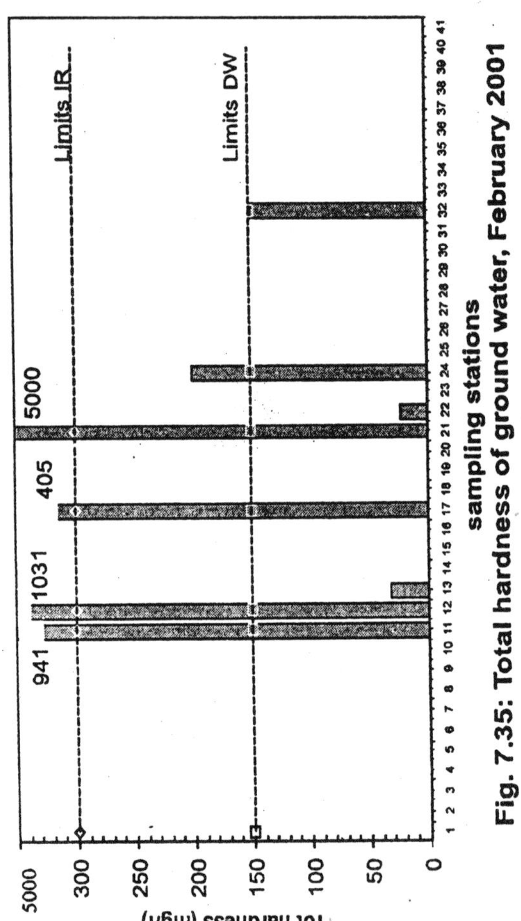

Fig. 7.35: Total hardness of ground water, February 2001

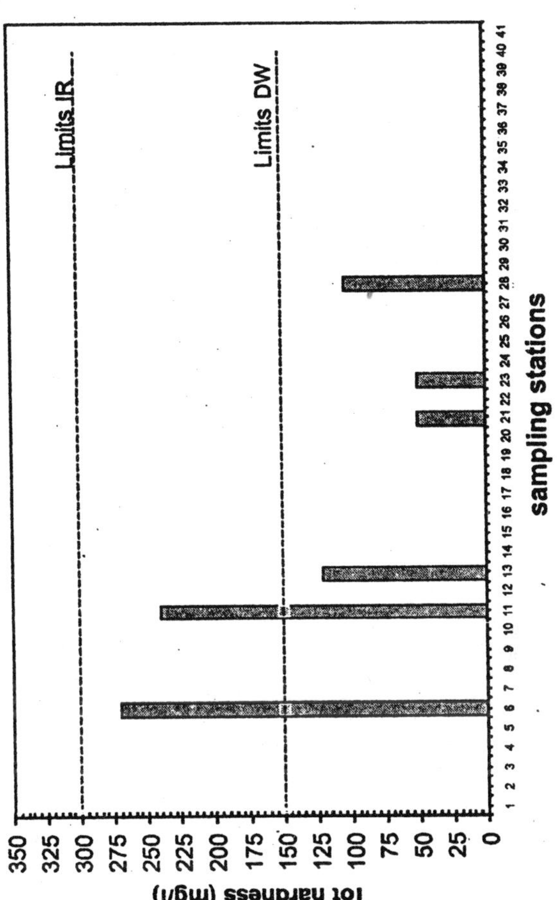

Fig. 7.36: Total hardness of ground water, April 2001

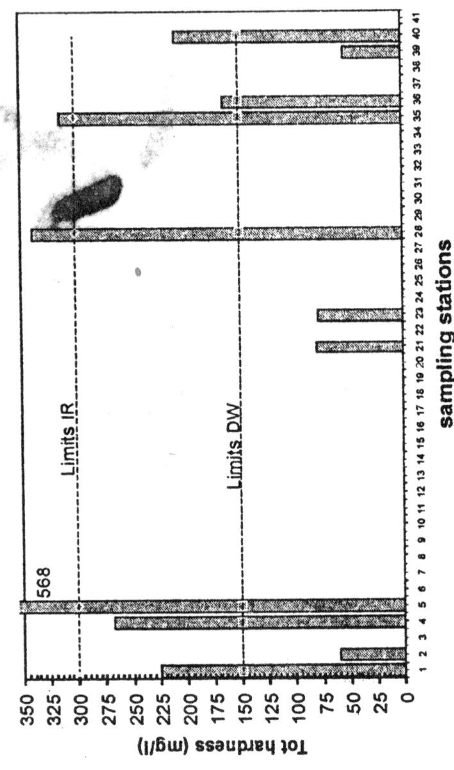

Fig. 7.37: Total hardness of ground water, June 2001

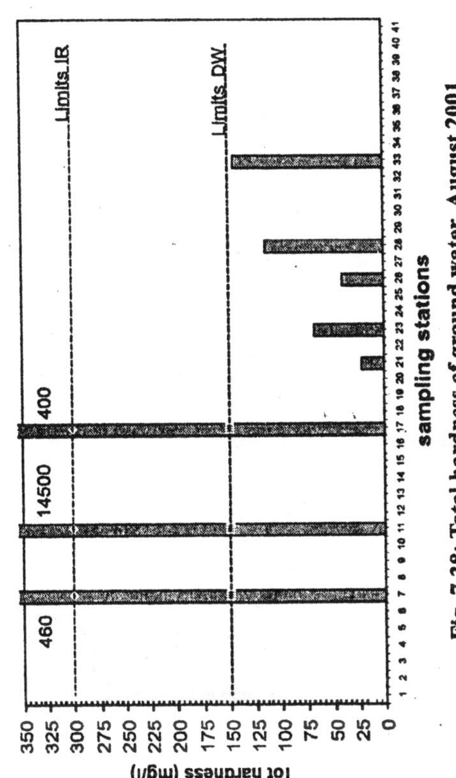

Fig. 7.38: Total hardness of ground water, August 2001

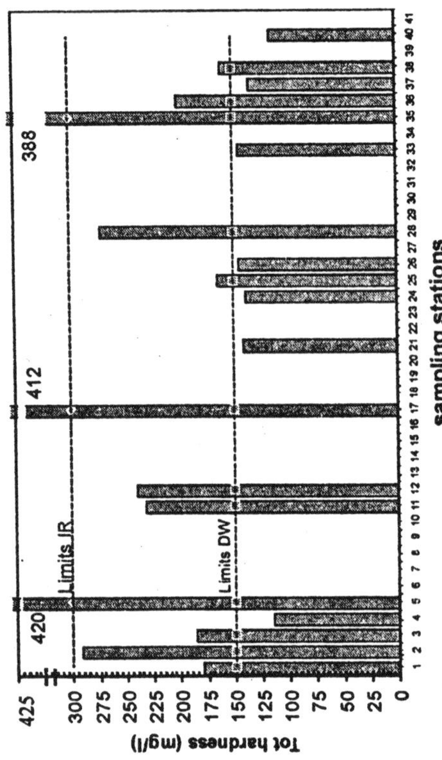

Fig. 7.39: Total hardness of ground water, October 2001

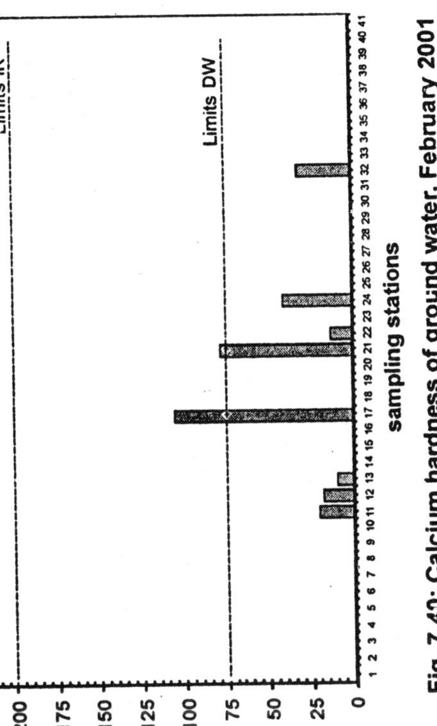

Fig. 7.40: Calcium hardness of ground water, February 2001

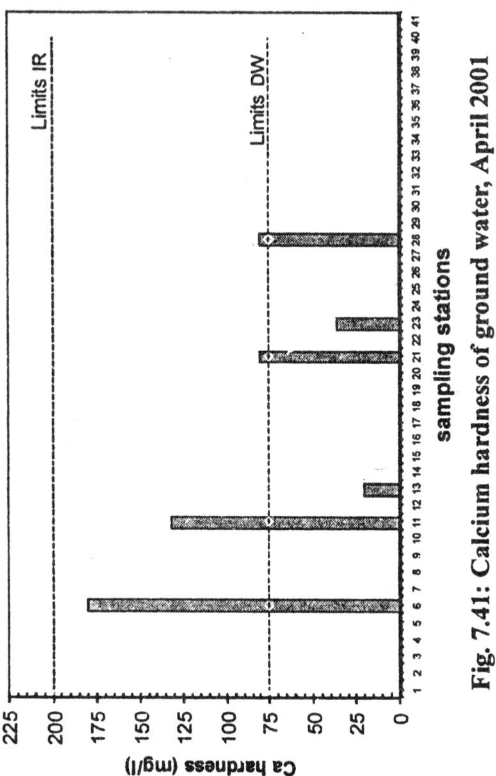

Fig. 7.41: Calcium hardness of ground water, April 2001

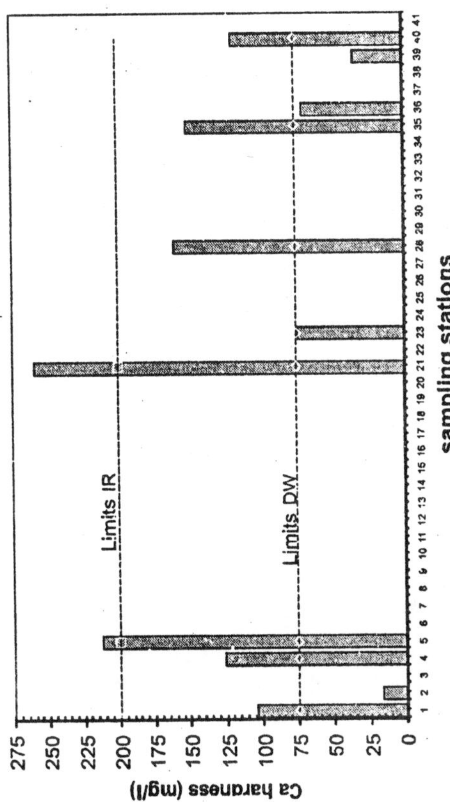

Fig. 7.42: Calcium hardness of ground water, June 2001

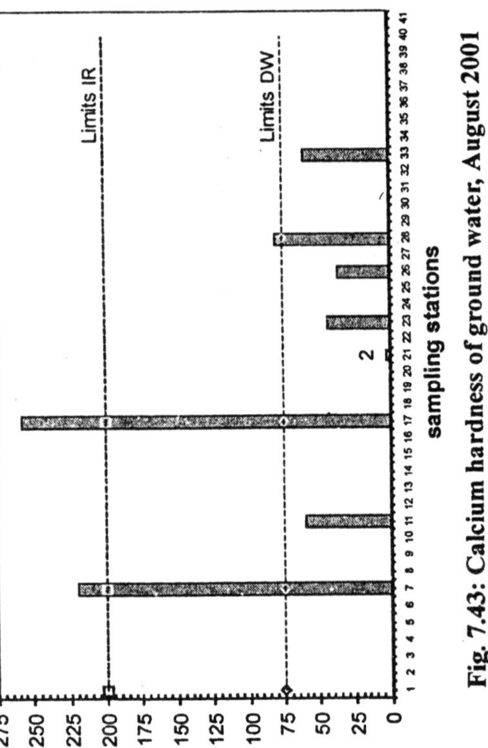

Fig. 7.43: Calcium hardness of ground water, August 2001

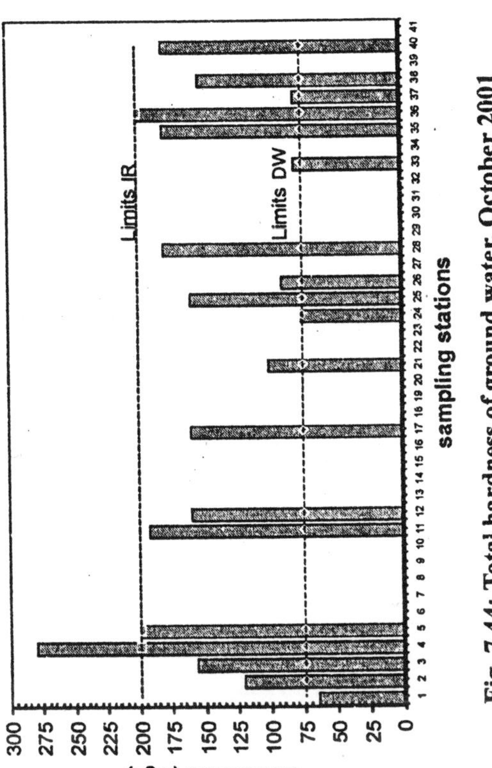

Fig. 7.44: Total hardness of ground water, October 2001

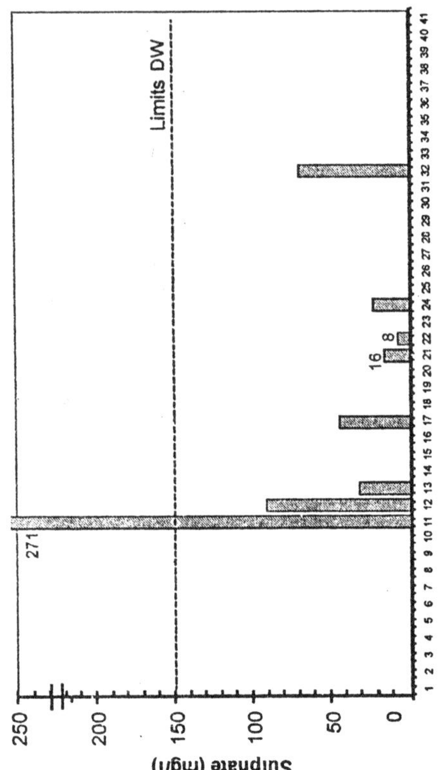

Fig. 7.45: Sulfates of ground water, February 2001

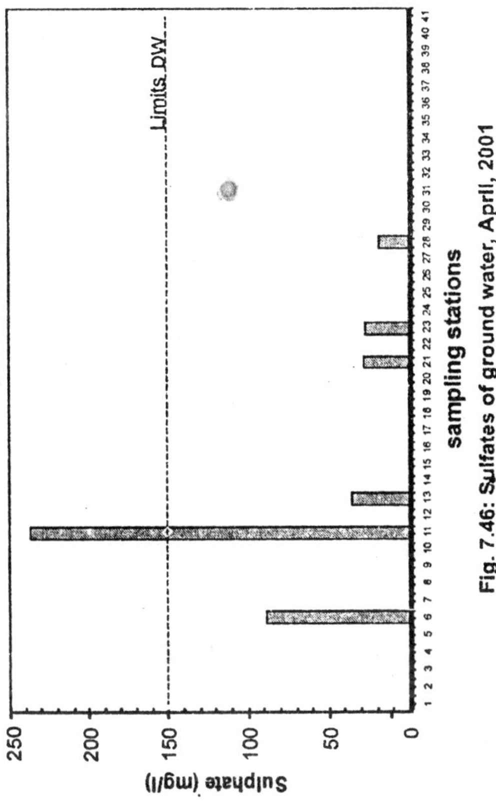

Fig. 7.46: Sulfates of ground water, April, 2001

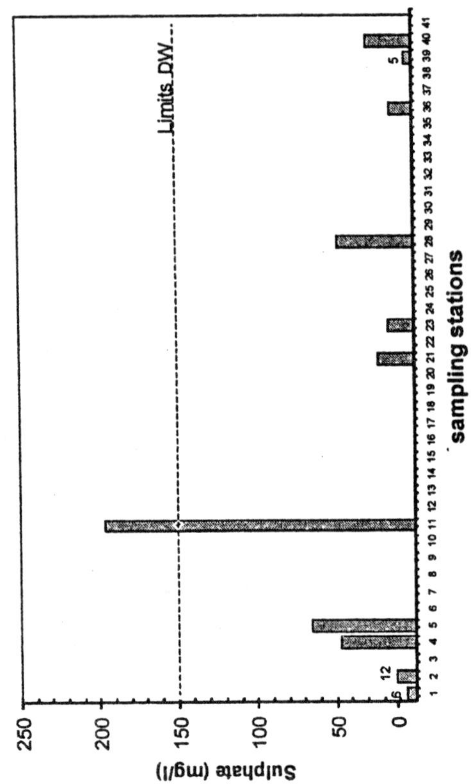

Fig. 7.47: Sulfates of ground water, June 2001

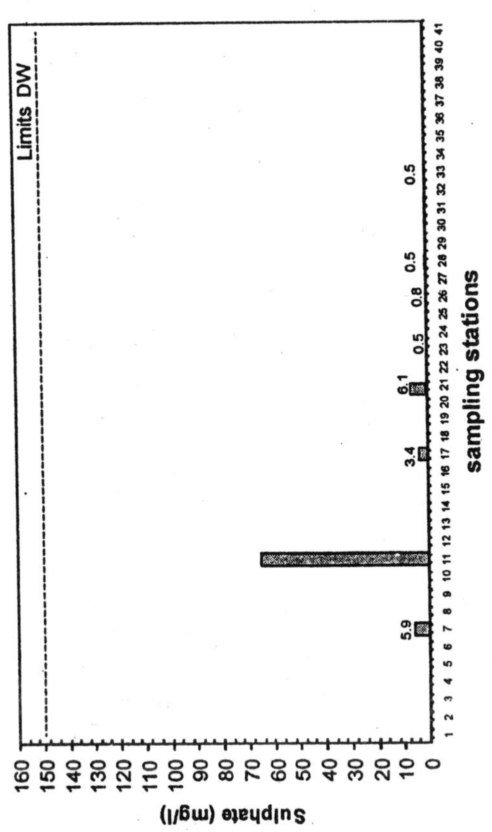

Fig. 7.48: Sulfates of ground water, August 2001

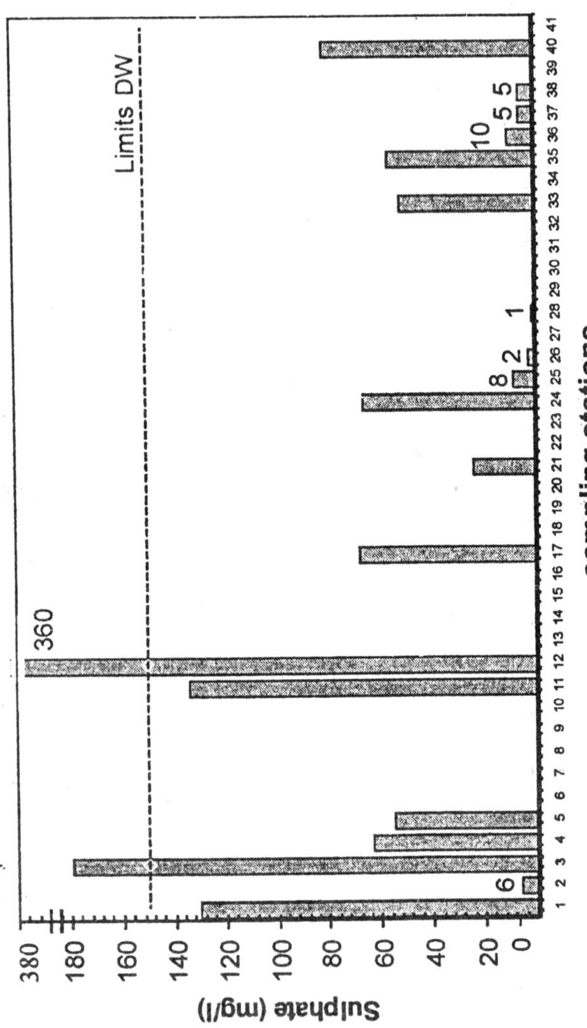

Fig. 7.49: Sulphate of ground water, October 2001

During the month of Aug '01, the sampling stations GW11 (Ferdous Nagar) and GW21 (Vadagaram) exceeded the permissible limits of drinking water quality criteria (BIS, 1991). However, rest of the months, both during monsoon and post monsoon, all the sampling sites were within the permissible limits.

Ammonical Nitrogen: During monsoon, ammonical nitrogen in the ground water ranged between 0.0057mg l^{-1} (GW21, Vadagaram, Jun'01) and 2.19 mg l^{-1} (GW24, Mudaliarpet, Oct '01) (Figs. 7.66-7.68).

Ammonical nitrogen in the ground water during post monsoon varied between 0.0787 mg l^{-1} (GW21, Vadagaram, Apl '01) and 0.5724 mg l^{-1} (GW21, Vadagaram, Feb '01) (Figs. 7.64, 7.65). Higher levels of ammonia was observed during peak monsoon.

Organic Nitrogen: The Organic nitrogen of the ground water varied between 0.23 mg l^{-1} (GW39, Velur, Jun '01) and 2.13 mg l^{-1} (GW3, Kodur, Oct' 01) during the monsoon (Figs. 7.71-7.73); and 0.3009 mg l^{-1} (GW17-Kilpettai, GW22-Kandadu, Feb '01) and 1.132 mg l^{-1} (GW13, Seyyankuppam, Apl' 01) (Figs. 7.69, 7.70) during post monsoon.

Phosphorus

Phosphate levels in the ground water ranged between 0.02 mg l^{-1} (GW4-Vivanattam, Oct'01) and 1.06 mg l^{-1} (GW17-Kilpettai, Oct '01) during monsoon (Figs. 7.76-7.78); 0.0017 mg l^{-1} (GW11-Ferdous Nagar, Feb '01) and 1.304 mg l^{-1} (GW12-Kunimedu, Feb '01) during post monsoon (Figures 7.74, 7.75).

During the month of April '01, a majority of the samples, 83%, were above the permissible limits of drinking (BIS, 1991) and irrigation water quality criteria (BIS, 1974). During monsoon 39% of the samples had phosphate levels above the permissible limits of drinking (BIS, 1991) and irrigation water quality criteria (BIS, 1974). However, during the post monsoon, samples that were above the permissible limits rose to 60.5%.

The high concentration of phosphorus in the ground water can be attributed to the following factors: (i) intensive agriculture practice in the catchment and (ii) the decomposing vegetation of the lake (iii) the high pH of the ground water at some of the sites can be responsible for leaching of soil bound phosphorus (Vanloon and Duffy, 2000).

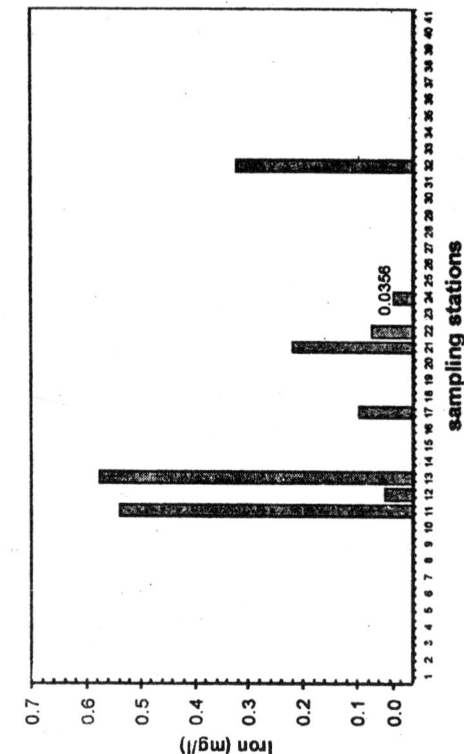

Fig. 7.50: Total iron of ground water, February 2001

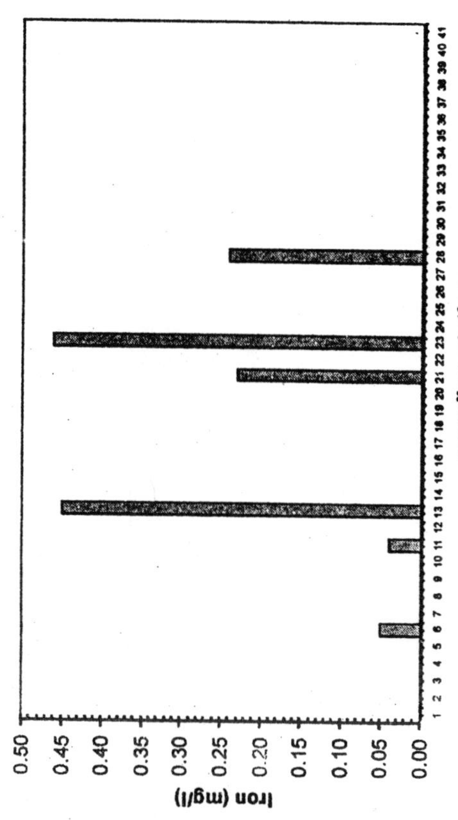

Fig. 7.51: Total iron of ground water, April 2001

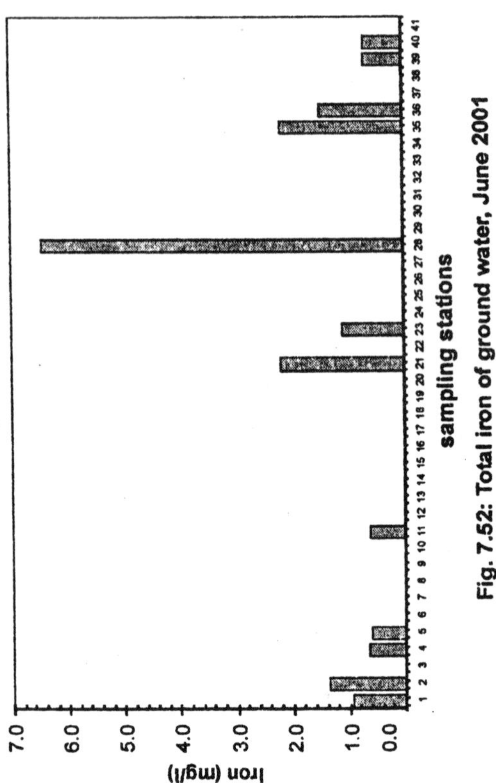

Fig. 7.52: Total iron of ground water, June 2001

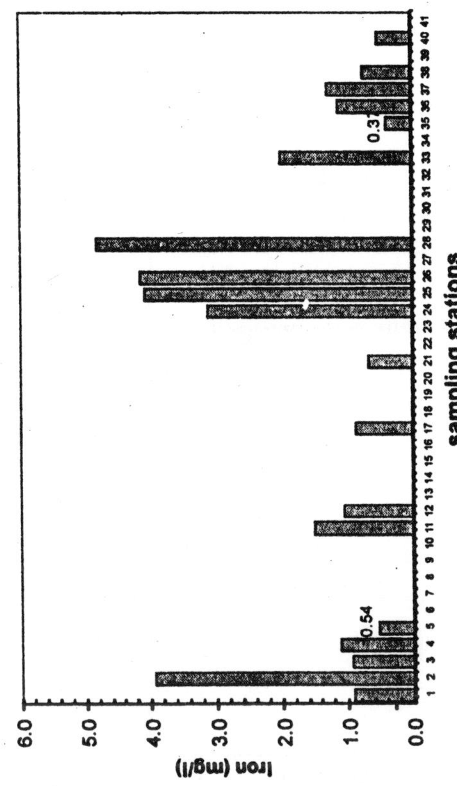

Fig. 7.53: Total iron of ground water, October 2001

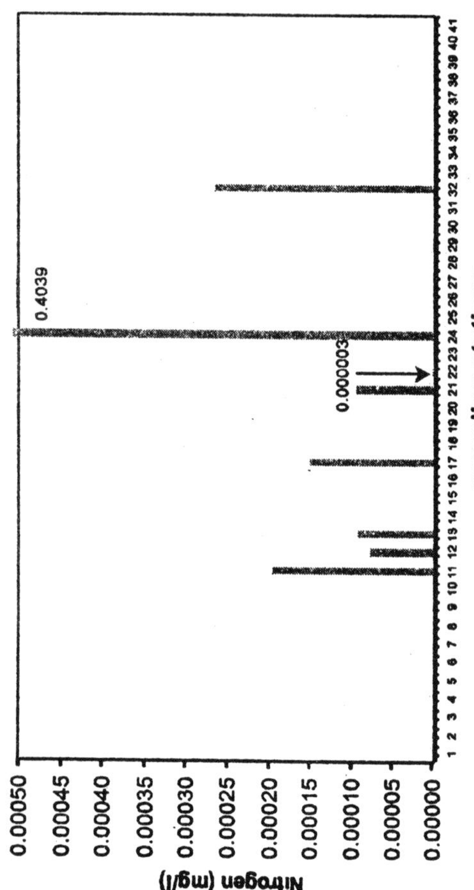

Fig. 7.54: Nitrites of ground water, February 2001

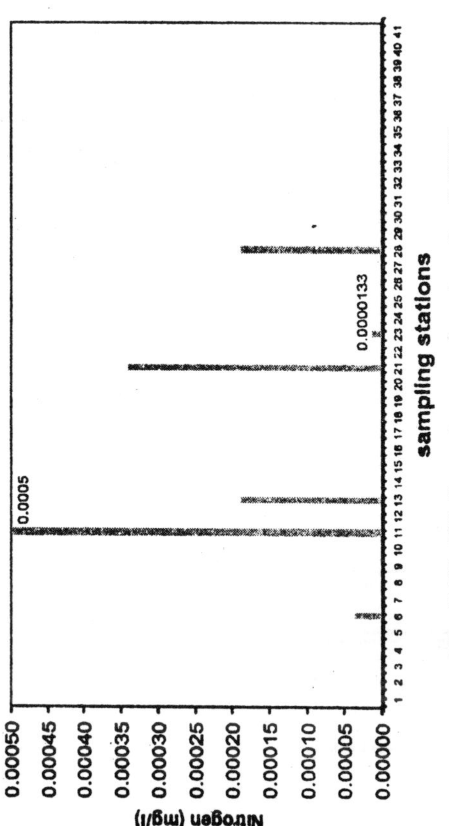

Fig. 7.55: Nitrites of ground water, April 2001

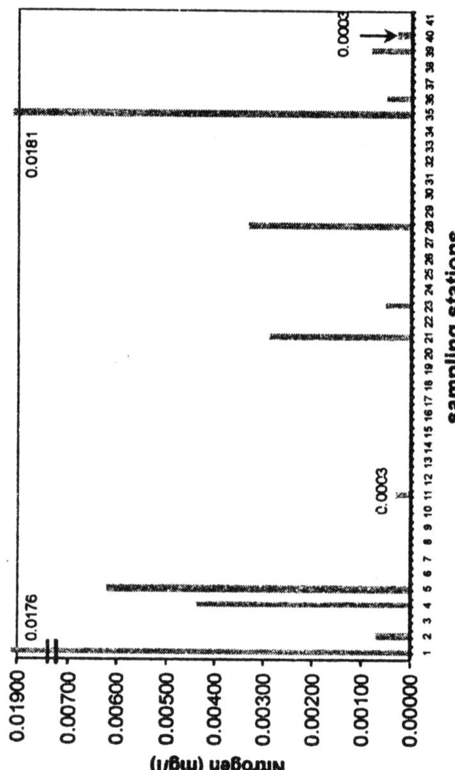

Fig. 7.56: Nitrites of ground water, June 2001

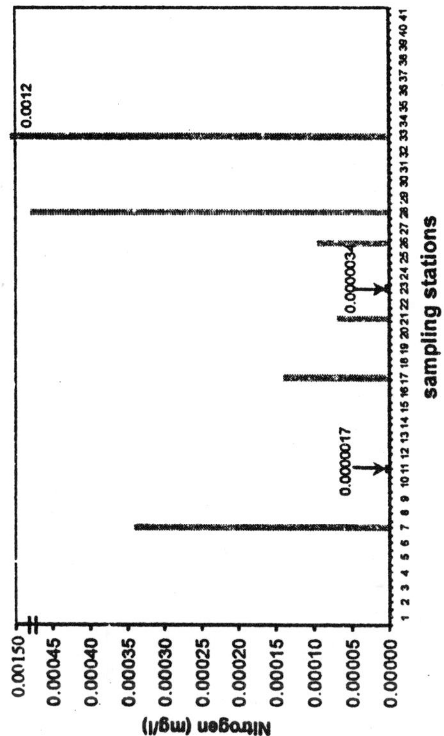

Fig. 7.57: Nitrites of ground water, August 2001

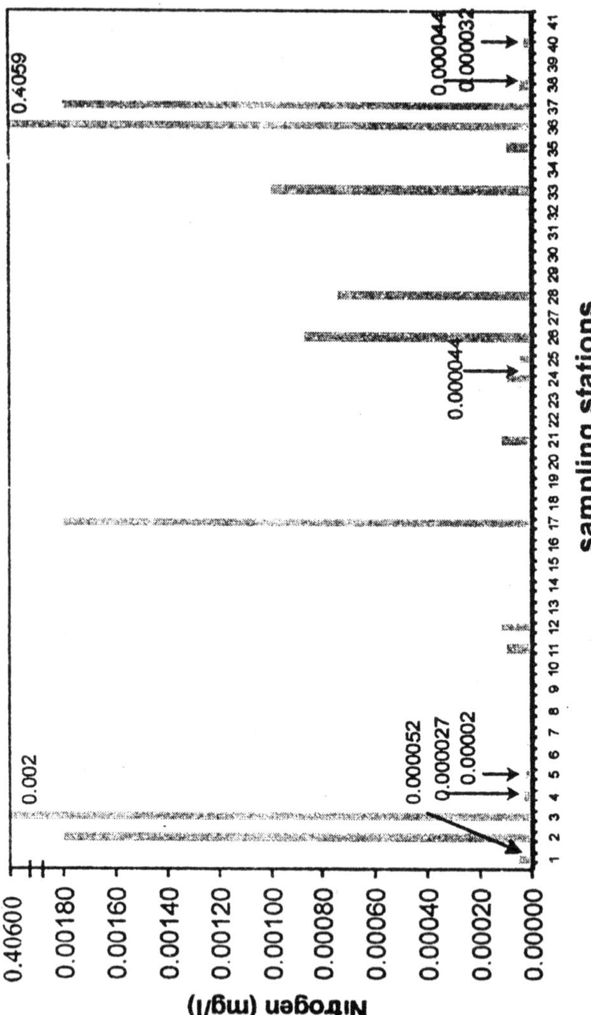

Fig. 7.58: Nitrites of ground water, October 2001

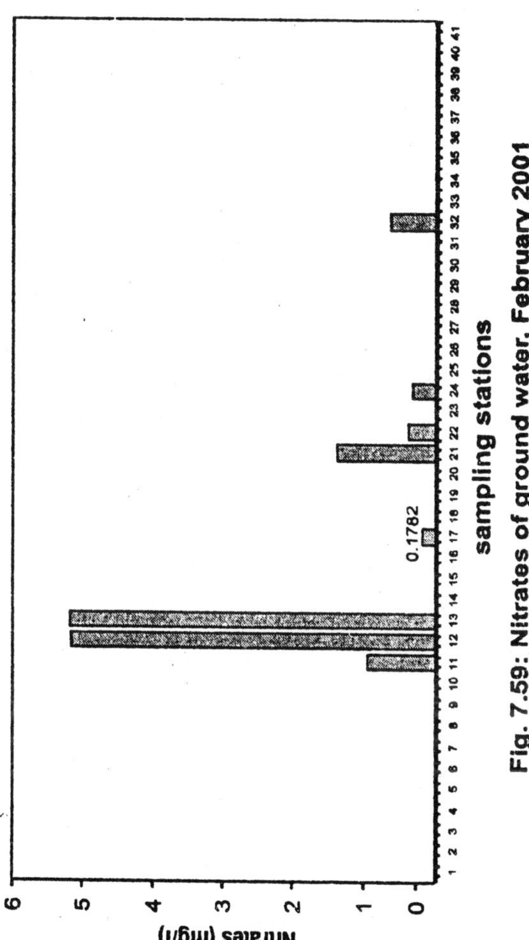

Fig. 7.59: Nitrates of ground water, February 2001

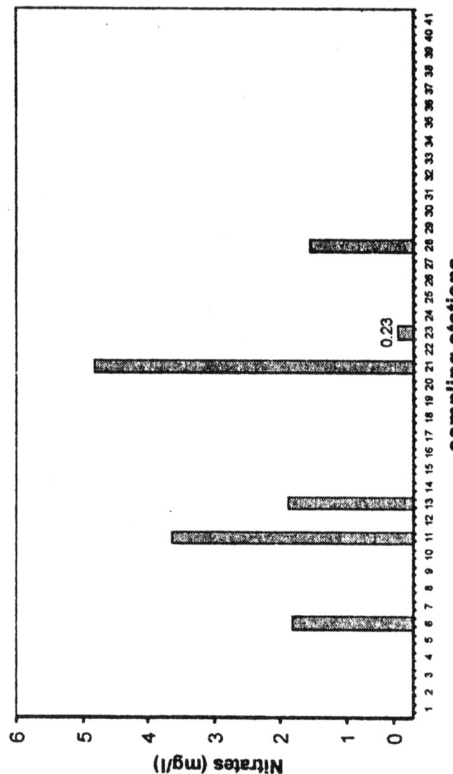

Fig. 7.60: Nitrates of ground water, April 2001

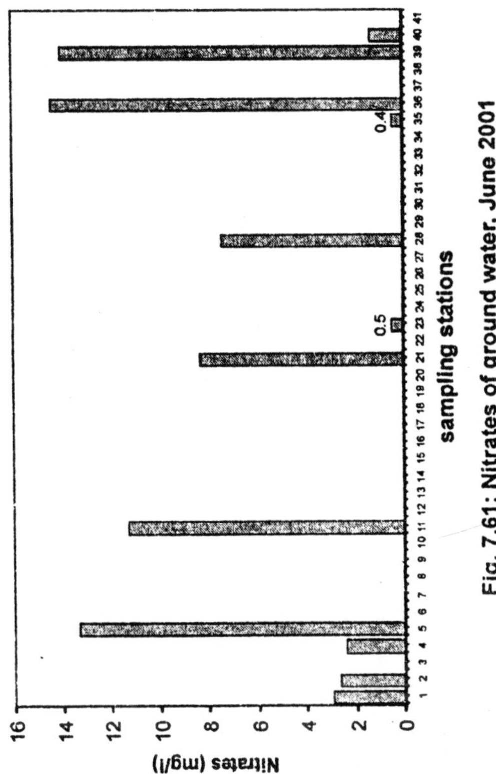

Fig. 7.61: Nitrates of ground water, June 2001

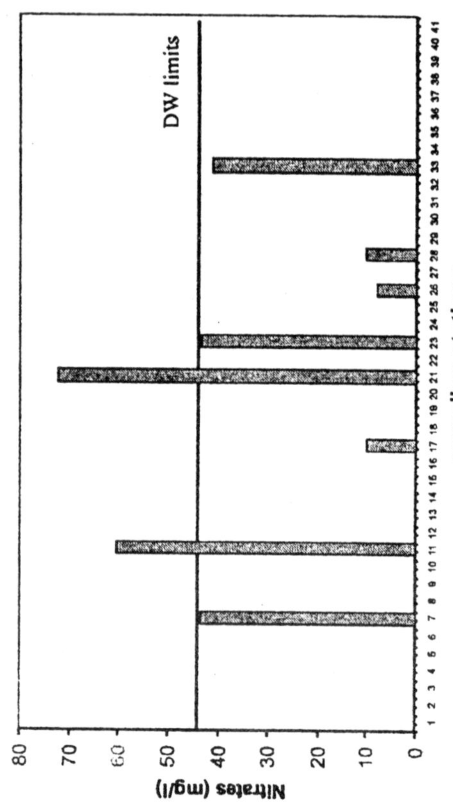

Fig. 7.62: Nitrates of ground water, August 2001

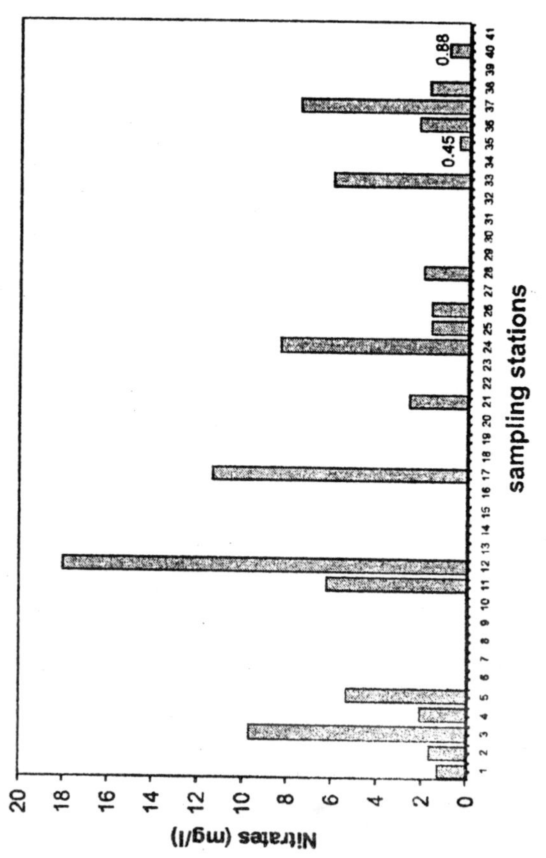

Fig. 7.63: Nitrates of ground water, October 2001

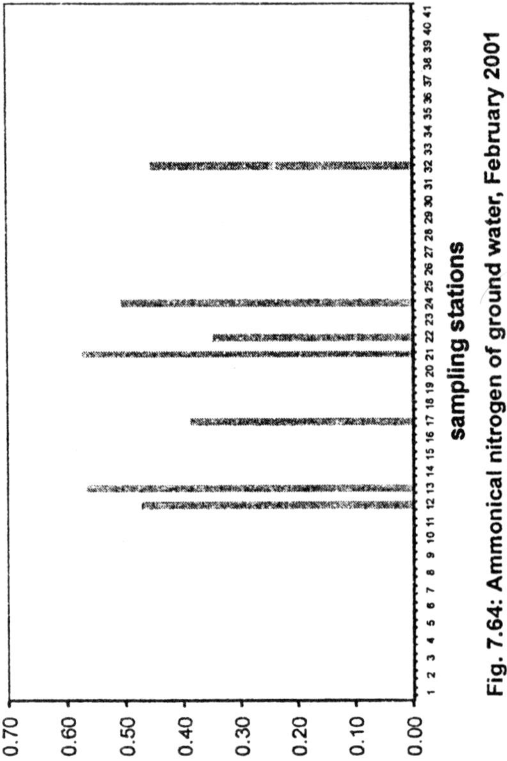

Fig. 7.64: Ammonical nitrogen of ground water, February 2001

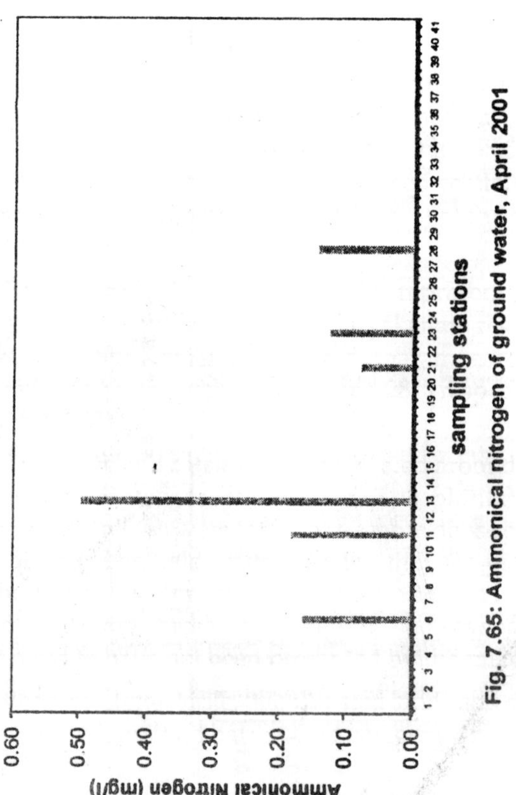

Fig. 7.65: Ammonical nitrogen of ground water, April 2001

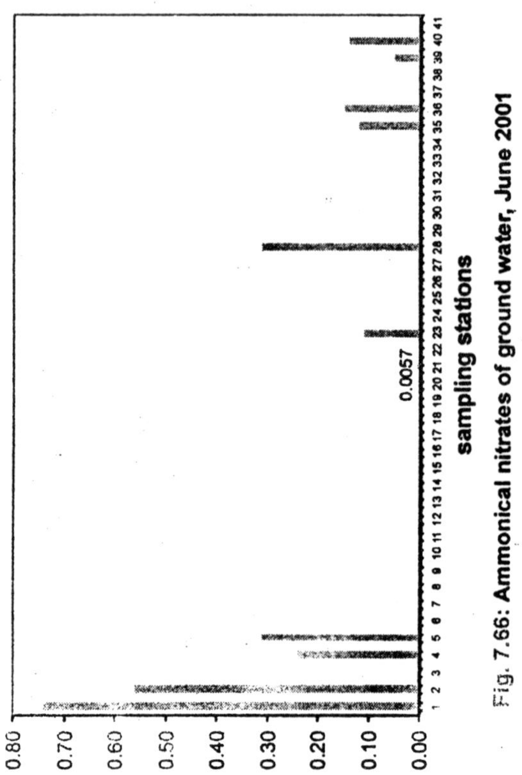

Fig. 7.66: Ammonical nitrates of ground water, June 2001

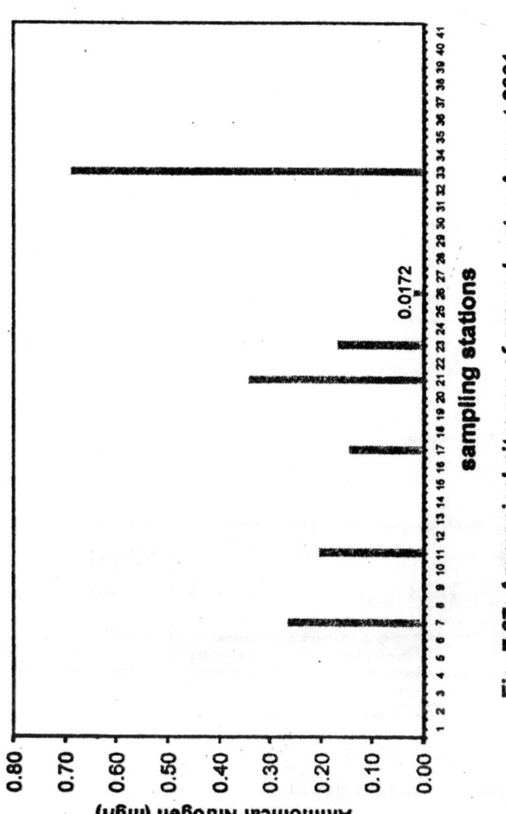

Fig. 7.67: Ammonical nitrogen of ground water, August 2001

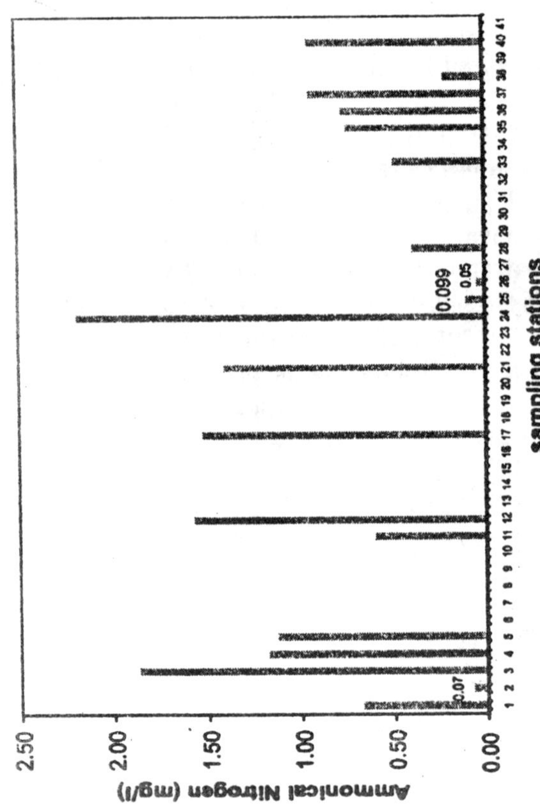

Fig. 7.68: Ammonical nitrogen of ground water, October 2001

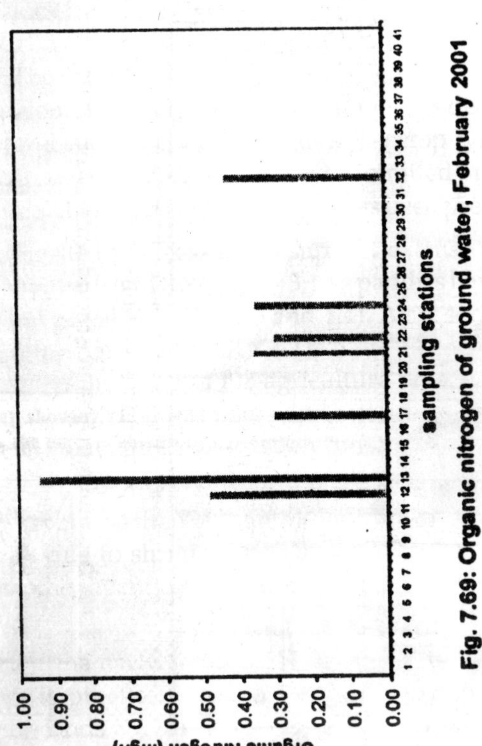

Fig. 7.69: Organic nitrogen of ground water, February 2001

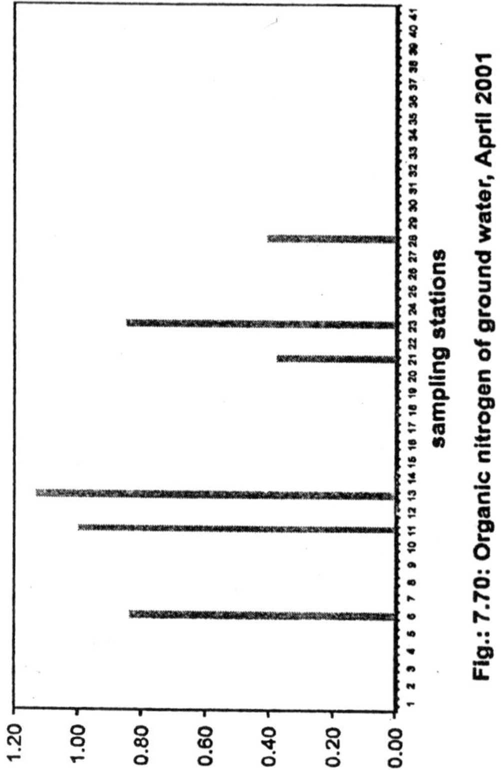

Fig.: 7.70: Organic nitrogen of ground water, April 2001

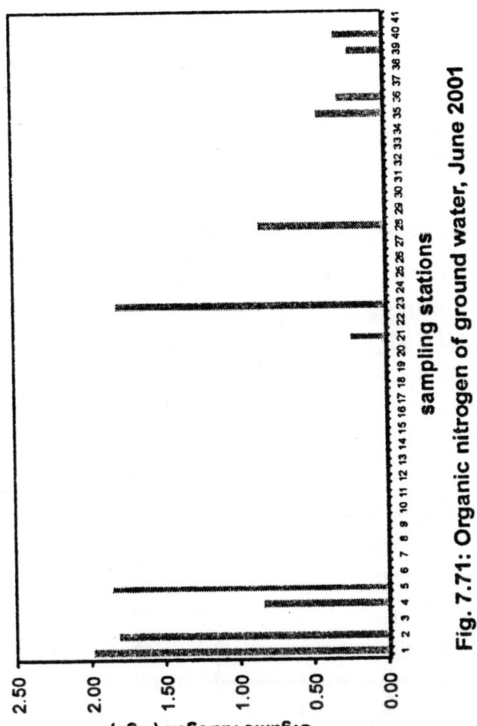

Fig. 7.71: Organic nitrogen of ground water, June 2001

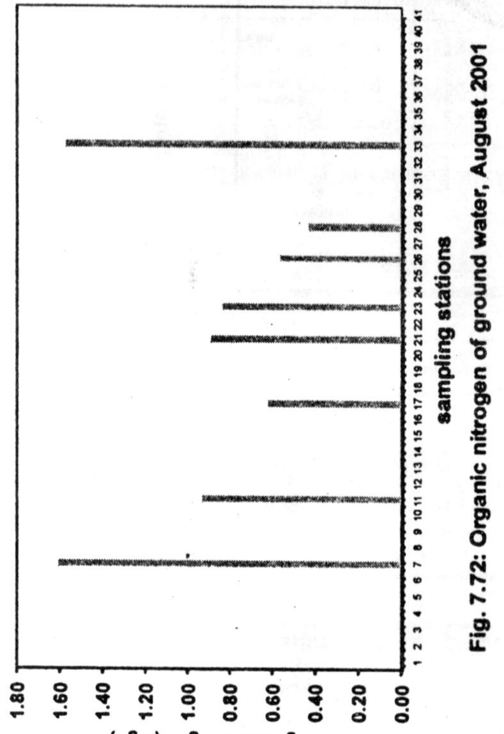

Fig. 7.72: Organic nitrogen of ground water, August 2001

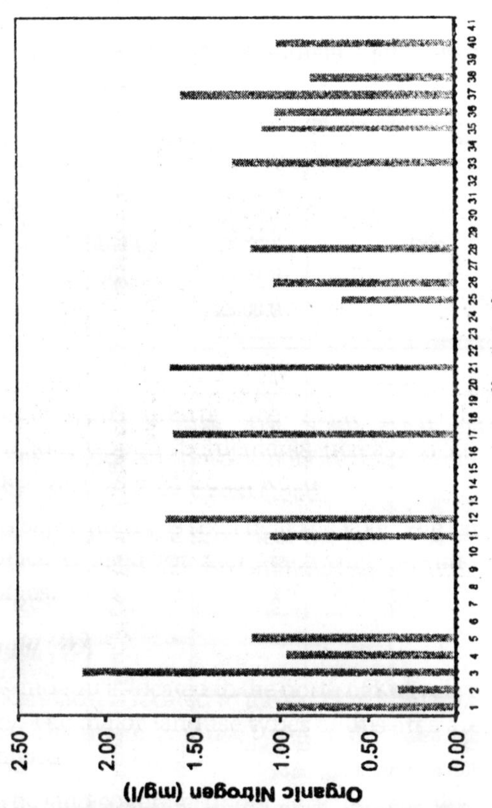

Fig. 7.73: Organic nitrogen of ground water, October 2001

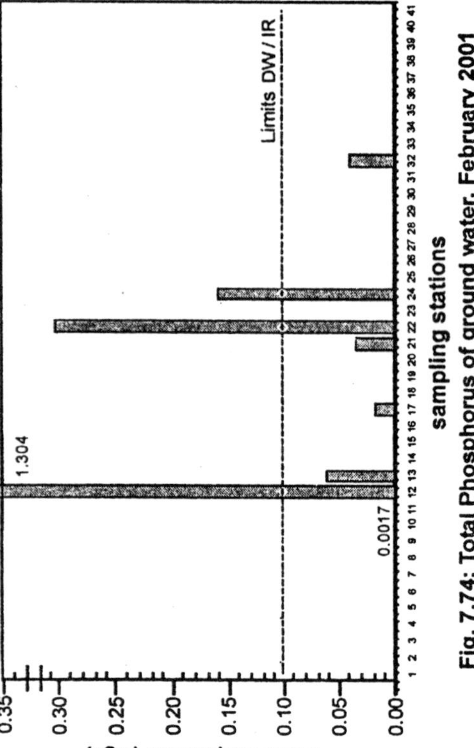

Fig. 7.74: Total Phosphorus of ground water, February 2001

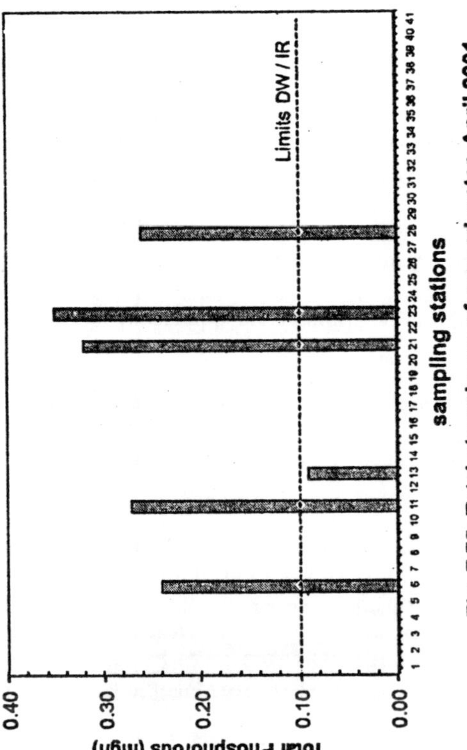

Fig. 7.75: Total phosphorus of ground water, April 2001

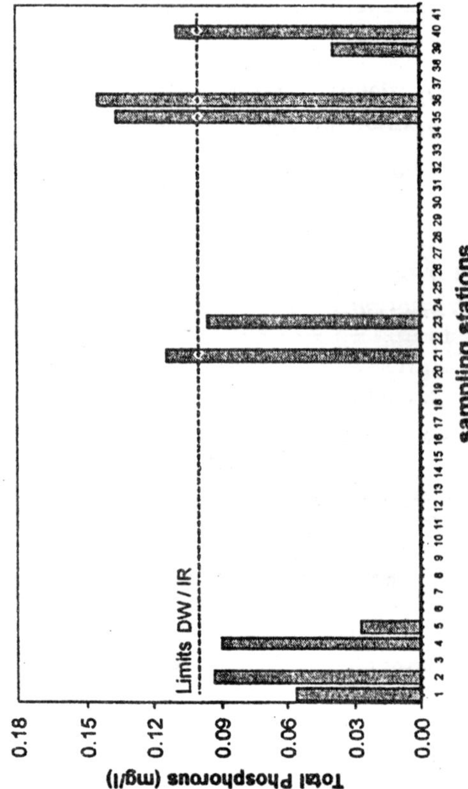

Fig. 7.76: Total phosphorus of ground water, June 2001

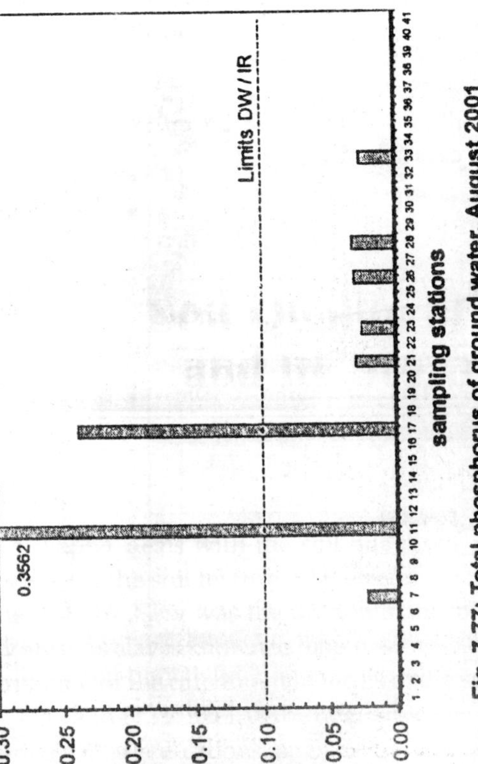

Fig. 7.77: Total phosphorus of ground water, August 2001

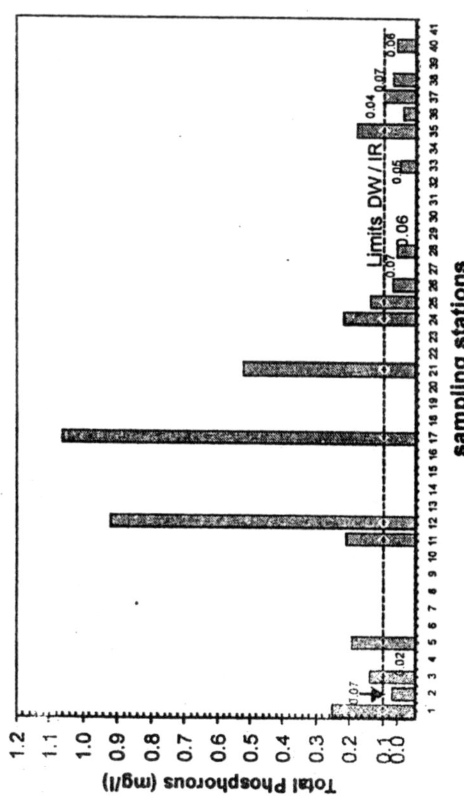

Fig. 7.78: Total phosphorus of ground water, October 2001

Conclusions

The present study on the ground water of Kaliveli reveals that precipitation, and the subsequent leaching of the soil constituents, and evapotranspiration are the primary factors that determine the ground water quality. The predominant chemical species that enter the ground water aquifers are nitrogen, phosphorus—where the soils are acidic, and the areas with intensive agriculture practice, and iron—at the areas where water is logged for long time, or , the ground water HRT had been for considerably long.

Kaliveli faces a serious threat of eutrophication because of the following factors: (i) landuse/land cover pattern as already mentioned in the chapter 14, (ii) the soil type, which is predominantly clay, can leach the nitrites very readily—both into the ground water aquifers and the lake as run-off, and (iii) the heavy drafting of the ground water for irrigation.

The imminent threat that Kaliveli region is now facing from the intensive practice of shrimp farming (Chari, 1997), which would not only aggravate the soil chemistry but also pollutes the ground and surface water.

In recent times there has been a rapid increase in urbanization and industrialization of Kaliveli region. Already, a pharmaceutical industry (SHASUN), an alkali manufacturing unit (Chemfab alkalies pvt. ltd), and Pondicherry University is located in this watershed. A recent addition to this cluster is a medical college and hospital (Pondicherry Institute of Medical Sciences). The present intensity of urbanization is so rapid that before one can realize, it would be too late to find the remedies that would plague the ground water of this region very soon. In this context it is imperative that sufficient care is taken not to over draft the ground water in this region. Rainwater harvesting and recharging of the ground water aquifers are some of the options which can alleviate the water quality of Kaliveli region.

REFERENCES

APHA, 1998 *Standard Methods for the Examination of Water and Waste Water*, Washington DC 20005.

BIS, 1974. *Indian Standard Quality for Irrigation Water*, Bureau of Indian Standards, Manak Bhavan, New Delhi.

BIS, 1994. *Indian Standard Quality Tolerances for Fresh Water for Fish Culture*, Bureau of Indian Standards, Manak Bhavan, New Delhi.

BIS, 1991, *Indian Standard Drinking Water-Specifications*, Bureau of Indian Standards, Manak Bhavan, New Delhi.

Chari, K.B. 1997. A Reconnaissance of the Ecology of Wetlands : Ousteri and Kaliveli, M.S. thesis, p. 110. Pondicherry University, India.

Sarwar S. G and Rifet, 1991. *Abiotic Functions of a Fresh Water Lotic Ecosystem of Kashmir (Dhoodh Ganga River)*. Poll.res.,Vol:10 (2), pp. 69-73.

Vanloon G.W., and Duffy S.J., 2000. *Environmental Chemistry: A Global Perspective*, Oxford University Press, p: 492.

ANNEXURE

Table—7.1: Abbreviations used for the various parameters in the subsequent tables and their units of measurement

	Abbreviations	*Units*
pH	pH	pH units
EC	Electrical conductivity	mmhos
Alk	Alkalinity	mgl^{-1}
Aci	Acidity	mgl^{-1}
CaH	Calcium hardness	mgl^{-1}
TotH	Total hardness	mgl^{-1}
Cl	Chlorides	mgl^{-1}
NO_2	Nitrites	mgl^{-1}
NO_3	Nitrates	mgl^{-1}
NH_3	Ammonical nitrogen	mgl^{-1}
OrgN	Organic nitrogen	mgl^{-1}
$RePO_4$	Reactive phosphorus	mgl^{-1}
$TotPO_4$	Total phosphorus	mgl^{-1}
SO_4	Sulfates	mgl^{-1}
Fe	Iron	mgl^{-1}

Table—7.2: Reconnaissance of groundwater quality in the vicinity of the Kaliveli wetland, March 2000

Sam No	*place*	*pH*	*EC (mmohs)*
GW6	Puthupattu	7.3	1.23
GW8	Puthupattu	7.0	0.91
GW10	Puthupattu	5.2	0.21
GW11	Ferdous Na	5.1	2.10
GW12	Kunimedu	6.2	0.79
GW13	Seyyankupp	5.6	0.32
GW14	Seyyankupp	6.1	0.39
GW15	Anumandai	5.8	3.43
GW16	Anumandai	6.9	0.43
GW17	Kilpettai	6.9	1.39
GW19	Kalikuppam	6.8	0.32
GW20	Kalikuppam	7.1	0.84
GW21	Vadagaram	5.2	1.93
GW22	Kandadu	5.8	0.38
GW25	Mudaliarpet	5.8	0.40
GW26	Pachapaith	6.2	0.62
GW28	Tirukanur	5.9	0.68
GW29	Nadukuppam	5.8	0.58
GW30	Vandipalay	5.4	0.12
GW31	Vandipalay	5.4	0.16
GW32	Vandipalay	6.1	0.56

Table—7.3: Reconnaissance of groundwater quality in the vicinity of Kaliveli wetland, April 2000

Sam No	*Place*	*pH*	*EC (mmhos)*
GW6	Puthupattu	7.6	3.38
GW9	Puthupattu	6.7	1.38
GW10	Puthupattu	5.1	0.24
GW11	Ferdous Na	4.4	0.99
GW12	Kunimedu	6.4	0.74
GW13	Seyyankupp	6.1	0.37
GW14	Seyyankupp	6.4	0.36
GW15	Anumandai	6.1	3.21
GW16	Anumandai	8.0	0.62
GW17	Kilpettai	7.1	1.42
GW19	Kalikuppam	7.0	0.34
GW20	Kalikuppam	7.3	0.88
GW21	Vadagaram	5.1	2.03
GW22	Kandadu	5.9	0.42
GW25	Mudaliarpet	5.8	0.45
GW28	Tirukanur	6.0	0.69
GW29	Nadukuppam	6.3	0.48
GW30	Vandipalay	5.2	0.14
GW31	Vandipalay	6.8	0.70
GW32	Vandipalay	6.4	0.72

Table—7.4: Reconnaissance of groundwater quality in the vicinity of Kaliveli wetland, June 2000

Sam No	*Place*	*pH*	*EC (mmohs)*
GW6	Puthupattu	7.2	3.52
GW8	Puthupattu	7.1	0.98
GW9	Puthupattu	7.3	0.82
GW10	Puthupattu	5.7	0.25
GW11	Ferdous Na	5.1	1.08
GW12	Kunimedu	6.5	1.01
GW13	Seyyankupp	6.1	0.33
GW14	Seyyankupp	6.4	0.43
GW15	Anumandai	6.2	3.74
GW16	Anumandai	7.0	0.78
GW19	Kalikuppam	7.2	0.38
GW20	Kalikuppam	7.8	0.93
GW21	Vadagaram	5.2	2.25
GW22	Kandadu	6.0	0.34
GW25	Mudaliarpet	7.0	0.51
GW28	Tirukanur	6.5	0.62
GW30	Vandipalay	6.4	1.91
GW31	Vandipalay	7.5	0.73
GW32	Vandipalay	7.1	0.72

Table—7.5: Reconnaissance of groundwater quality in the vicinity of the Kaliveli wetland, February 2001

Sam No	*Place*	*pH*	*EC (mmohs)*
GW6	Puthupattu	6.8	2.0
GW7	Puthupattu	6.7	1.50
GW8	Puthupattu	6.7	0.82
GW10	Puthupattu	5.0	0.15
GW11	Ferdous Na	5.0	1.20
GW12	Kunimedu	6.0	0.25
GW13	Seyyankupp	6.3	0.25
GW14	Seyyankupp	6.7	0.40
GW15	Anumandai	5.7	2.52
GW17	Kilpettai	6.7	1.10
GW19	Kalikuppam	6.6	0.26
GW20	Kalikuppam	6.9	0.69
GW21	Vadagaram	4.5	1.68
GW22	Kandadu	5.3	0.18
GW24	Mudaliarpet	5.9	0.61
GW28	Tirukanur	5.8	0.47
GW29	Nadukuppam	6.1	0.34
GW32	Vandipalay	6.1	0.88

Table—7.6: Ground water quality in the vicinity of the Kaliveli wetland, February 2001

Sam No	place	pH	EC (mmohs)	Alk	Aci	CaH	TotH	Chl	NO_2	NO_3	NH_3	OrgN	$TotPO_4$	SO_4	Fe
GW11	Ferdous Na	5.0	1.20	12	120	20	941	181	0.000196	0.9504	-	–	0.0017	271.07	0.54
GW12	Kunimedu	6.0	0.25	12	40	18	1031	102	0.000078	5.1732	0.4729	0.4794	1.304	91.19	0.05
GW13	Seyyankupp	6.3	0.25	8.0	40	9.0	31	46	0.0000915	5.1894	0.5662	0.9435	0.0602	31.64	0.5752
GW17	Kilpettai	6.7	1.10	48	200	105	405	2339	0.00015	0.1782	0.3858	0.3009	0.0172	44.08	0.0981
GW21	Vadagaram	4.5	1.68	12	160	78	5000	86	0.0000945	1.3824	0.5724	0.357	0.0344	15.82	0.2186
GW22	Kandadu	5.3	0.18	16	100	13	22	62	0.000003	0.378	0.3484	0.3009	0.3027	7.821	0.073
GW24	Mudaliarpet	5.9	0.61	28	120	41	200	54	0.4039	0.324	0.504	0.3519	0.1582	22.57	0.0356
GW32	Vandipalay	6.1	0.88	12	120	31	149	134	0.0002655	0.621	0.4542	0.4335	0.0395	69.68	0.3218

Table—7.7: Reconnaissance of groundwater quality in the vicinity of the Kaliveli wetland, April 2001

Sam No	*Place*	*PH*	*EC (mmhos)*
GW6	Puthupattu	7.2	2.93
GW9	Puthupattu	7.0	1.20
GW10	Puthupattu	5.0	0.30
GW11	Ferdous Na	5.4	1.43
GW13	Seyyankupp	6.9	0.35
GW14	Seyyankupp	6.9	0.82
GW15	Anumandai	7.2	3.72
GW19	Kalikuppam	6.8	0.38
GW20	Kalikuppam	7.3	1.02
GW21	Vadagaram	5.6	2.30
GW23	Mudaliarpet	6.0	0.13
GW28	Tirukanur	6.0	0.53

Table—7.8: Groundwater quality in the vicinity of the Kaliveli wetland, April 2001

Sam No	*Place*	*PH*	*EC (mmhos)*	*Alk*	*Aci*	*CaH*	*TotH*	*Chl*	NO_2	NO_3	NH_3	*OrgN*	*TotPO*$_4$	SO_4	*Fe*
GW6	Puthupattu	7.2	2.93	40	20	180	270	333	0.000038	1.83	0.1633	0.838	0.24	88.4	0.05
GW11	Ferdous Na	5.4	1.43	40	60	132	240	138	0.0005	3.63	0.182	0.998	0.27	236.7	0.04
GW13	Seyyankupp	6.9	0.35	60	20	20	120	62	0.00019	1.88	0.4993	1.132	0.09	35.28	0.45
GW21	Vadagaram	5.6	2.30	100	40	80	50	311	0.00034	4.84	0.0787	0.379	0.32	27.89	0.23
GW23	Mudaliarpet	6.0	0.13	40	20	36	50	62	1.33E-05	0.23	0.126	0.849	0.35	27.28	0.46
GW28	Tirukanur	6.0	0.53	80	60	80	104	124	0.00019	1.55	0.1446	0.409	0.26	19.07	0.24

Table—7.9: Reconnaissance of groundwater qualit: in the vicinity of the Kaliveli wetlands, June 2001

Sam No	*Place*	*PH*	*EC (mmohs)*
GW1	Ichchangad	7.6	0.74
GW2	Anpakkam	7.7	0.80
GW3	Kodur	7.8	1.35
GW4	Vilvanatta	7.2	1.43
GW5	Kaluperumb	7.3	1.32
GW6	Puthupattu	7.5	0.24
GW7	Puthupattu	7.2	1.17
GW10	Puthupattu	5.6	0.25
GW11	Ferdous Na	4.1	1.54
GW15	Anumandai	6.6	3.31
GW19	Kalikuppam	7.5	0.43
GW20	Kalikuppam	7.2	0.98
GW21	Vadagaram	6.2	2.45
GW23	Mudaliarpet	5.8	0.25
GW24	Mudaliarpet	6.1	0.58
GW28	Tirukanur	6.8	1.40
GW31	Vandipalay	6.8	0.54
GW34	Talaganiku	7.3	1.07
GW35	Talaganiku	6.9	1.03
GW36	Talaganiku	7.5	0.12
GW37	Devanandal	6.9	0.47
GW38	Katterimed	8.0	0.61
GW39	Velur	6.1	0.44
GW40	Pudukuppam	7.5	1.08
GW41	Dhanamaniy	7.3	1.72

Table—7.10: Groundwater quality in the vicinity of the Kaliveli wetlands, June 2001

Sam No	Place	PH	EC (mmohs)	Alk	Aci	CaH	TotH	Chl	NO_2	NO_3	NH_3	OrgN	$RePO_4$	$TotPO_4$	SO_4	Fe
GW1	Ichchangad	7.6	0.74	300	40	104	224	40	0.02	2.9	0.74	1.99	0.02	0.056	5.51	0.94
GW2	Anpakkam	7.7	0.80	340	16	16	60	20	0	2.6	0.56	1.81	0.08	0.093	12.27	1.36
GW4	Vilvanatta	7.2	1.43	400	20	126	268	170	0	2.4	0.24	0.8385	0.073	0.09	46.62	0.65
GW5	Kaluperumb	7.3	1.32	480	28	212	568	190	0.01	13.3	0.31	1.8546	-	0.027	65.93	0.6
GW11	Ferdous Na	4.1	1.54	60	20	-	-	250	0	11.3	-	-	0.0025	-	196.8	0.63
GW21	Vadagaram	6.2	2.45	80	24	260	80	550	0	8.3	0.006	0.2359	0.0175	0.114	22.89	2.19
GW23	Mudaliarpet	5.8	0.25	180	12	74	78	15	0	0.5	0.11	1.8126	0.029	0.096	15.99	1.09
GW28	Tirukanur	6.8	1.40	100	20	160	340	150	0	7.4	0.31	0.8523	0.0737	-	47.58	6.45
GW35	Talaganiku	6.9	1.03	240	40	150	314	152	0.02	0.4	0.12	0.4508	-	0.136	-	2.17
GW36	Talaganiku	7.5	0.12	220	12	70	164	38	0	14.4	0.15	0.31	-	0.144	14.06	1.46
GW39	Velur	6.1	0.44	60	16	34	54	175	0	14	0.05	0.23	0.0175	0.039	4.68	0.67
GW40	Pudukuppam	7.5	1.08	280	16	118	208	114	0	1.33	0.14	0.33	0.05	0.109	28.27	0.69

Table—7.11: Reconnaissance of groundwater quality in the vicinity of the Kaliveli wetland, August 2001

Sam No	*Place*	*pH*	*EC (mmhos)*
GW6	Puthupattu	7.2	2.00
GW7	Puthụpattu	7.4	0.89
GW11	Ferdous Na	4.1	1.08
GW15	Anumandai	6.4	2.22
GW17	Kilpettai	6.9	1.17
GW18	Kilpettai	6.8	1.05
GW19	Kalikuppam	6.9	0.45
GW20	Kalikuppam	8.0	0.90
GW21	Vadagaram	6.0	1.62
GW23	Mudaliarpet	7.4	0.71
GW24	Mudaliarpet	7.3	0.75
GW26	Pachapaith	6.3	0.61
GW27	Pudupakkam	6.4	0.42
GW28	Tirukanur	6.4	0.62
GW30	Vandipalay	6.1	0.42
GW31	Vandipalay	6.5	0.62
GW33	Vandipalay	6.9	1.80

Table—7.12: Groundwater quality in the vicinity of the Kaliveli wetland, August 2001

Sam No	Place	pH	EC (mmhos)	Alk	Aci	CaH	TotH	Chl	NO_2	NO_3	NH_3	OrgN	$RePO_4$	$TotPO_4$	SO_4
GW7	Puthupattu	7.4	0.89	51	28	220	460	102	0.00034	43.521	0.266	1.6038	-	0.022	5.86
GW11	Ferdous Na	4.1	1.08	5	38	60	14500	23042	1.7E-06	60.416	0.203	0.9272	0.0062	0.3562	65.2
GW17	Kilpettai	6.9	1.17	100	30	258	400	146	0.00014	10.208	0.146	0.6247	0.0221	0.2377	3.39
GW21	Vadagaram	6.0	1.62	16	45	2	22	461	6.8E-05	72.4	0.343	0.892	-	0.0295	6.06
GW23	Mudaliarpet	7.4	0.71	13	48	43	68	32	3.4E-06	43.282	0.166	0.834	0.0012	0.0247	0.52
GW26	Pachapaith	6.3	0.61	14	13	36	40	28	9.5E-05	8.059	0.017	0.5684	-	0.0304	0.79
GW28	Tirukanur	6.4	0.62	34	16	80	114	47	0.00048	10.507		0.4348	-	0.0312	0.48
GW33	Vandipalay	6.9	1.80	52	12	60	144	593	0.0012	41.252	0.687	1.5744	-	0.0255	0.54

Table—7.13: Reconnaissance of groundwater quality in the vicinity of the Kaliveli wetland, October 2001

Sam No	*Place*	*PH*	*EC (mmhos)*
GW1	Ichchangad	6.8	1.68
GW2	Anpakkam	7.0	0.98
GW3	Kodur	7.1	1.74
GW4	Vilvanatta	6.7	1.42
GW5	Kaluperumb	6.8	1.51
GW6	Puthupattu	7.1	2.85
GW7	Puthupattu	6.8	1.10
GW9	Puthupattu	6.8	0.93
GW11	Ferdous Na	3.5	1.26
GW12	Kunimedu	6.3	0.41
GW15	Anumandai	7.8	2.58
GW17	Kilpettai	6.6	1.51
GW18	Kilpettai	6.3	1.04
GW19	Kalikuppam	6.8	0.44
GW20	Kalikuppam	7.0	0.82
GW21	Vadagaram	5.3	1.90
GW24	Mudaliarpet	6.3	0.67
GW25	Mudaliarpet	5.7	0.46
GW26	Pachapaith	5.4	0.23
GW27	Pudupakkam	5.7	0.26
GW28	Tirukanur	5.8	0.46
GW30	Vandipalay	5.6	0.31
GW33	Vandipalay	6.5	1.47
GW34	Talaganiku	7.0	0.89
GW35	Talaganiku	6.2	1.01
GW36	Talaganiku	6.8	0.68
GW38	Katterimed	5.4	0.35
GW39	Velur	7.0	0.93
GW40	Pudukuppam	6.7	1.53
GW41	Dhanamaniy	6.7	2.09

Table—7.14: Groundwater quality in the vicinity of the Kaliveli wetland, October 2001

Sam No	Place	PH	EC (mmhos)	Alk	Aci	CaH	TotH	Chl	NO2	NO3	NH3	OrgN	TotPO4	SO4	Fe
GW1	Ichchangad	6.8	1.68	61	77	65	180	330	0.000052	1.26	0.68	1.02	0.25	130	0.92
GW2	Anpakkam	7.0	0.98	88	160	120	292	96	0.0018	1.66	0.07	0.33	0.07	6	3.94
GW3	Kodur	7.1	1.74	300	220	156	186	296	0.002	9.67	1.87	2.13	0.14	179	0.95
GW4	Vilvanatta	6.7	1.42	170	130	280	114	191	0.000027	2.1	1.18	0.97	0.02	63	1.11
GW5	Kaluperumb	6.8	1.51	105	60	200	420	232	0.00002	5.33	1.13	1.16	0.19	55	0.54
GW11	Ferdous Na	3.5	1.26	72	138	192	232	212	0.000093	6.24	0.6	1.06	0.21	134	1.49
GW12	Kunimedu	6.3	0.41	182	145	160	240	46	0.00012	17.98	1.57	1.65	0.92	360	1.06
GW17	Kilpettai	6.6	1.51	2	125	160	412	232	0.0018	11.33	1.52	1.61	1.06	68	0.87
GW21	Vadagaram	5.3	1.90	107	110	100	140	572	0.00012	2.58	1.4	1.63	0.52	24	0.68
GW24	Mudaliarpet	6.3	0.67	210	78	76	138	907	0.0001	8.35	2.19	-	0.22	66	3.11
GW25	Mudaliarpet	5.7	0.46	95	135	160	164	122	0.000044	1.62	0.099	0.65	0.14	8	4.07
GW26	Pachapaith	5.4	0.23	67	142	90	144	56	0.00087	1.66	0.05	1.03	0.07	2	4.15
GW28	Tirukanur	5.8	0.46	90	162	180	272	102	0.00074	2.04	0.39	1.16	0.06	1	4.82
GW33	Vandipalay	6.5	1.47	138	90	80	144	335	0.001	6.05	0.49	1.26	0.05	51	2
GW35	Talaganiku	6.2	1.01	46	56	180	388	203	0.000093	0.45	0.74	1.1	0.18	56	0.37
GW36	Talaganiku	6.8	0.68	122	120	200	200	93	0.4059	2.22	0.77	1.02	0.04	10	1.11
GW37	Devanandal	-	-	295	208	80	134	96	0.0018	7.52	0.94	1.56	0.1	5	1.28
GW38	Katterimed	5.4	0.35	76	67	152	160	76	0.000044	1.76	0.21	0.83	0.07	5	0.75
GW40	Pudukuppam	6.7	1.53	58	46	180	114	232	0.000032	0.88	0.95	1.01	0.06	80	0.52

SECTION—V

PRODUCTIVITY, FLORA AND FAUNA

8

Primary Productivity of Kaliveli
*A Rare Estuarine Wetland of South Indian Peninsula**

Abstract

To assess the environmental status of Kaliveli wetland, in situ monitoring of primary productivity and other key water quality parameters was done in the Kaliveli wetland. The Yedayanthitu estuary, which connects Kaliveli to the Bay of Bengal, was also studied.

The average gross primary productivity and the net primary productivity of Kaliveli wetland was slightly higher than the Yedayanthitu estuary, by 23.22 mg C m^{-3} h^{-1} and 27 mg C m^{-3} h^{-1} respectively. However, the respiration was higher in the Yedayanthitu estuary by 3.57 mg C m^{-3} h^{-1}. However, no significant difference was observed in the water quality of the wetland and the estuary in terms of pH, electrical conductivity, and dissolved oxygen.

---

* *Reproduced from the Proceedings of 'The Conference of Hydraulics, Water Resources and Ocean Engineering-HYDRO 2002', pp. 497-501.*

A comparison of primary productivity of Kaliveli with other lakes of India reveals that the former is tending towards eutrophy.

Key words: Estuary, wetland, primary productivity, water quality, eutrophication.

Introduction

Wetlands, such as Kaliveli, are extremely important in harvesting rainwater, recharging ground water, fisheries, and in numerous other ways. In order to ensure sustainability of such wetlands, it is imperative that the wetlands are periodically monitored *vis-à-vis* their environmental health, and preventive steps are taken whenever any weakness in the health is diagnosed.

Measurement of primary productivity can be a very important diagnostic tool for a wetland, as important as measuring heartbeat is for a human being. This single parameter is capable of indicating the sum-total impact of dozens of influencing factors. Otherwise, for the evaluation of the state of natural waters the entire spectrum of physiological and biological interactions of the aquatic community must be considered (APHA, 1998).

Primary productivity of an aquatic ecosystem provides a comprehensive assessment of the trophic status, whether a given water body is oligotrophic, mesotrophic or eutrophic. While the estimation of different physico-chemical characteristics of water bodies reveals alteration in quality induced by the incoming pollutants, the estimation of primary productivity reveals alteration in its physiological status induced by the same (Shaw et al., 1991).

Although there is considerable amount of literature on primary production in temperate water bodies, relatively fewer studies have been reported on water bodies in the tropics, even less so in India (Vijayaraghavan, 1971; Srinivasan, 1976; Lal, 1989; Kumar, 1996; Harikrishnan and Azis, 2000). In view of all these reasons, the present study of the primary productivity of Kaliveli wetland was conducted.

Experimental Methods

Double-distilled and deionized water and analytical reagent grade chemicals were used for all analytical work. The procedures described in standard texts (APHA, 1998, Abbasi, 1998) were employed for all estimates. The pH, electrical conductivity, and dissolved oxygen were measured in situ employing Hanna pH electronic paper, Naina digital field conductivity meter, and Systronics dissolved oxygen meter, respectively.

The study was conducted in the month of November, when the lake is usually full due to the southwest monsoons (June to September) and the passing northeast monsoons (October to January). The study was carried out in triplicates at two sites, one in the Kaliveli wetland, at Anumandaikuppam, and the other in the Yedayanthitu estuary, at Erapannai (Fig. 8.1).

Primary productivity and respiration were measured using the light and dark bottle method of Gaardner and Gran (1927) as described in APHA (1998). The chief advantage of this method is that it provides estimates of gross and net productivity, and respiration. Water samples were collected in transparent and opaque bottles each of 300 ml capacity and exposed in situ at a depth of 30 cm for a photo-period of 7 hours from morning 6.30 am till 1.30 pm. After the completion of the incubation period, dissolved oxygen was estimated following Winkler's method (APHA, 1998). The oxygen values obtained were converted into the equivalent carbon values and the rate of production and estimation were expressed as mg C m^{-3} h^{-1}.

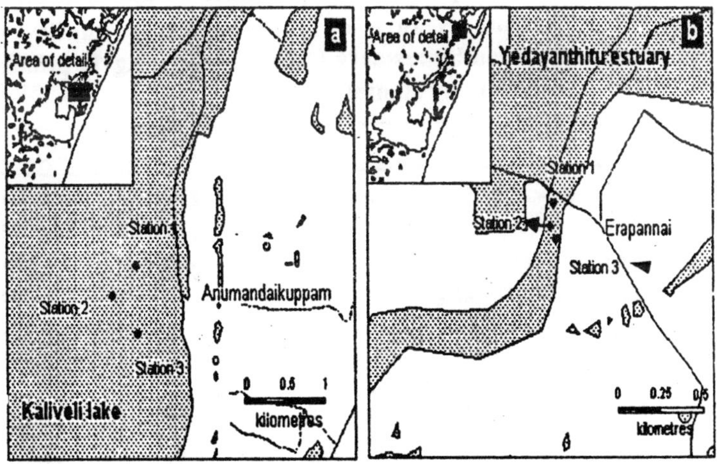

Fig. 8.1: Location of sampling stations from primary productivity study (a) Anumandaikuppam (b) Erapannai

Results and Discussion

During the study period the temperature of the water varied between 26°C (6.00 hrs) and 32°C (Table 8.1, Fig. 8.2a) in the Kaliveli wetland, at Anumandaikuppam; and 29°C (0600 hrs) and 32°C (0800 hrs) at Yedayanthitu estuary, near Erappannai (Table 8.2, Fig. 8.2b). The pH varied between 8.1 (0800 hrs) and 9.5 (1200 hrs) in the wetland (Fig. 8.2a), and 8.2 (0800 hrs) and 9.5 (1100 and 1200 hrs) in the estuary (Fig. 8.2b).

Table—8.1: Reconnaissance of the surface water quality of Kaliveli wetland at Anumandaikuppam during the primary productivity studies

Hrs	*Temp*	*PH*	*EC*	*DO*
0600	26.0	8.7	1.38	5.4
0700	26.5	8.4	1.13	3.5
0800	27.0	8.1	0.91	4.7
0900	29.0	8.8	0.62	7.6
1000	29.0	8.7	0.65	8.1
1100	29.5	9.4	1.19	10.0
1200	32.0	9.5	1.06	7.9
1300	32.0	8.6	0.65	9.1

Table–8.2: Reconnaissance of surface water quality at Yedayanthitu estuary at Erapannai during the primary productivity studies

Hrs	*Temp*	*pH*	*EC*	*DO*
0600	29.0	9.0	1.64	5.0
0700	31.0	8.5	1.02	4.8
0800	32.0	8.2	0.99	4.7
0900	30.5	9.0	1.66	7.5
1000	31.0	8.8	1.69	8.0
1100	31.0	9.5	1.56	8.0
1200	30.5	9.5	1.08	7.5
1300	31.0	8.6	0.89	9.0

Nov 2001; Cloud amount (visual interpretation): 85%

Figures 2a and 2b reveal that the dissolved oxygen (DO) values in the estuary were slightly less than what observed in the wetland. The lowest DO was observed at 0800 hrs (3.48 ppm) and the highest at 1100 hrs (10.00 ppm) in the wetland and in the estuary the lowest DO value was observed at 0800 hrs (4.7 ppm) and the highest at 1000 and 1100 hrs (8.00 ppm). The lower DO values in the estuary could be attributed to the high salinity which decreases the solubility of DO in water (Abbasi, 1998).

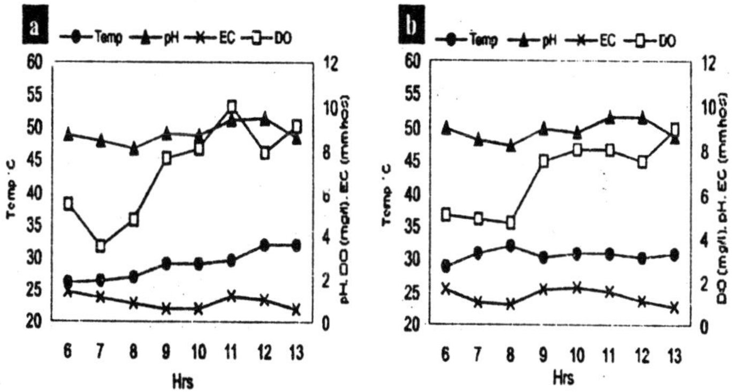

Fig. 8.2: Diurnal variance of temperature, pH, EC, and DO at (a) Anumandaikuppam, the fresh part of Kaliveli and (b) Erapannai, the estuarine part of Kaliveli

The electrical conductivity (EC) values in the wetland varied between 0.62 mmhos (0900 hrs) and 1.38 (0600 hrs). In the estuary the minimum EC value observed was 0.89 mmhos (1300 hrs) and the maximum 1.69 mmhos (1000 hrs).

The gross primary productivity (P_g) of the wetland varied between 202.50 mg C m^{-3} h^{-1} (station 2) and 315.54 mg C m^{-3} h^{-1} (station 1), with an average of 273.57 mg C m^{-3} h^{-1} (Table 8.3). In the estuary the gross primary productivity (P_g) varied between 217.50 mg C m^{-3} h^{-1} (station 1) and 267.56 mg C m^{-3} h^{-1} (station 2) with an average of 250.35 mg C m^{-3} h^{-1}.

The net primary productivity (P_n) values of the Kaliveli wetland varied between 159.64 mg C m^{-3} h^{-1} (station 2) and 267.32 mg C m^{-3} h^{-1} (station 1) with an average of 229 mg C m^{-3} h^{-1} (Table 8.3, Fig. 8.3a). The net primary productivity values in the estuary varied between 180 mg C m^{-3} h^{-1} (station 1) and 214.29 mg C m^{-3} h^{-1} (station 3) with an average of 202 mg C m^{-3} h^{-1} (Figure 8.3b).

Table—8.3 Primary productivity of Kaliveli wetland and Yedayanthitu estuary during the study period

Productivity	*Kaliveli wetland (Anumandaikuppam)*			*Yedayanthitu estuary (Erapannai)*		
	Station 1	*Station 2*	*Station 3*	*Station 1*	*Station 2*	*Station 3*
Pg ($m_gCm^{-3}h^{-1}$)	315.54	202.50	302.68	217.50	265.71	267.86
Pn ($m_gCm^{-3}h^{-1}$)	267.32	159.64	259.82	180.00	212.29	214.29
r ($m_gCm^{-3}h^{-1}$)	48.21	42.86	42.86	37.50	53.57	53.57

P_g-gross primary productivity; P_n-net primary productivity; r-respiration

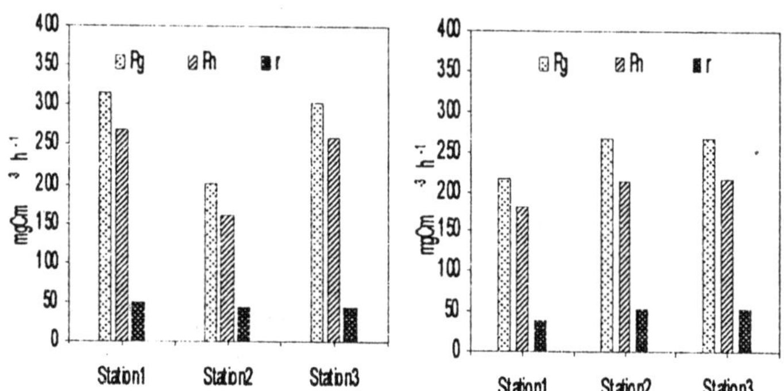

Fig. 8.3: The primary productivity values of Kaliveli wetland at (a) Anumandaikuppam and (b) Yedayanthitu estuary at Erapannai

Likens (1975) has considered nutrients as the primary controlling factor of primary productivity. The nutrient concentration normally limits the growth and production of phytoplankton (Williams, 1972). Kaliveli is rich in phosphorus and nitrogen (Selvi, 2001). The abundance of these nutrients would have contributed to the high growth of phytoplankton and hence the high productivity. Ketchen et. al. (1968) state that the ratio between net and gross productivity (P_n: P_g) should approach unity in a healthy population if the respiration is 5% to 10%. In the case of Kaliveli the ratio between net and gross productivity lies between 0.79% and 0.86% (Table 8.4) which is less than 1, indicating that the lake is Polluted (Mahajan and Kanhere, 1996).

Table—8.4: A comparison of the primary productivity of Kaliveli wetland and Yedayanthitu estuary with the other lakes of India

Lakes of India		P_g	P_n	r	% r (% of P_g)	P_n: P_g	P_n: r
Kaliveli (T.N) ($mgCm^{-3}h^{-1}$)	Anumandai	202-315	159-267	43-48	14-21	0.79-0.86	3.73-6.06
	Erapannai	218-268	180-214	38-53	17-20	0.80-0.83	3.96-4.80
Dobdia pond, (M.P)[1] ($mgCm^{-3}h^{-1}$)		112-250	75-262	37-112	22-50	0.78-0.55	1.22-3.5
Artificial pond (M.P)[2] ($mgCm^{-3}h^{-1}$)		75-31	50-237	25-87	24-43	0.57-0.77	1.33-3.10
Nani Barwani pond (M.P)[3] ($mgCm^{-3}h^{-1}$)		112-312	75-237	50 -100	24- 47	0.55-0.76	1.12-3.16
Hoogly estuary (W.B)[4] ($mgCm^{-3}h^{-1}$)		23-28	–	–	–	–	–
Rishimangalam tank (Kerala)[5] ($mgCm^{-3}day^{-1}$)		216-2843	72-1188	108-2135	50-75	0.27-0.52	0.66-0.56

P_g-gross primary productivity; Pn-net primary productivity; r-respiration

[1, 2, 3]Mahajan and Kanhere,1996; 4 Mitra et. al., 1994; 5 Harikrishnan and Azis, 2000

The respiration rate (r) of the plankton during the study varied between 42.86 mg C m^{-1} h^{-1} (station 2 and 3) in the wetland; and 37.50 mg C m^{-3} h^{-1} (station 1) and 53.57 mg C m^{-3} h^{-1} (station 2 and 3) in the estuary (Table 8.4). The net primary production (P_n) to the respiration (r) ratio exceeds 1 in the case of all the sampling stations which indicate that Kaliveli is polluted. Similar values of high P_n: r ratios have been reported for a tropical pond at Darbhanga (Ahmed, 1990) and for Yashwant Sagar reservoir (Sharma, 1993), but these two water bodies are much deeper than Kaliveli. The high P_n: r ratios in Kaliveli has dangerous portents because a shallow lake such as Kaliveli can silt-up much faster if it gets eutrophic. The higher respiration rate of the water in Kaliveli may be due to the decomposition of necrotic organic matter.

In general, the average gross primary productivity in the wetland (Anumandaikuppam) (Fig. 8.4) was slightly higher by 23.22 mg C m^{-3} h^{-1} than the estuary, while the net primary productivity and respiration were higher in the estuary by 6.94 and 3.57 mg C m^{-3} h^{-1} respectively. However, no significant difference was observed in the average gross and net primary productivity and respiration rate of plankton in the fresh water and the estuarine part of the lake.

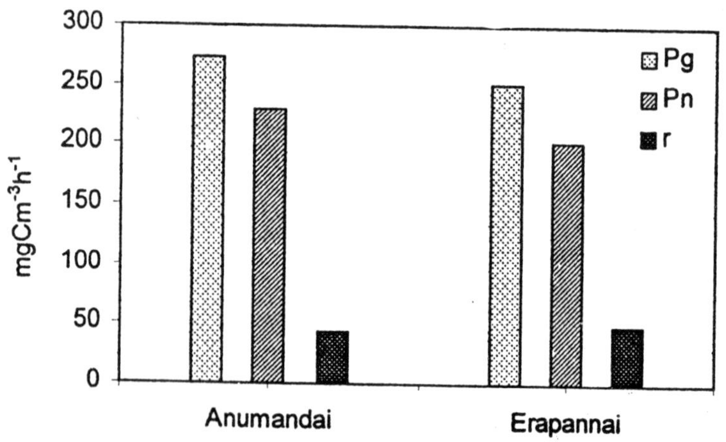

Fig. 8.4: Comparison of the average primary productivity values at Anumandaikuppam (Kaliveli Wetland) and Erapannai (Yedayanthitu estuary)

A comparison of primary productivity of Kaliveli with other lakes of India (Table 8.4) also reveals that the former is tending towards eutrophy.

REFERENCES

Abbasi, S.A., 1998. *Environmental Pollution and Its Control*, Cogent International, India, p. 439.

Ahmed, S.S., 1990. *Primary Production in a Tropical Pond at Darbhanga*, J.Nat.Con., pp. 1:1-6.

APHA, 1998. *Standard Methods for the Examination of Water and Wastewater*, American Public Health Association, Washington, DC.

Gaardner T., and Gran H.H., 1927. *Investigations on the Production of Plankton in the Oslo Fjord*, Rapp.P.V. Cons. Inst. Explor. Mer., pp. 144: 56-60.

Harikrishnan K., and Azis A., 2000. *Primary Production in a Freshwater Temple Tank in Kerala*, Indian J. Environ. & Ecoplan., pp. 3 (1): 127-130.

Ketchen B.H., Ryther J. H., Yentsch C.S., and Corwin N., 1968. *Productivity in Relation to Nutrients*, Rapp. P.V. Reun.Cons.Perm.Int.Explor. Mer., pp. 144: 132-140.

Kumar A., 1996. *Impact of Sewage Pollution on Chemistry and Primary Productivity of Two Fresh Water Bodies in Santal Pragana, India*, Indian J. Ecol., pp. 23: 86-92.

Lal A.K., 1989. *Seasonal Variation in Primary Productivity and Photosynthetic Efficiency in Tropical Swamps*, J.Fresh Wat. Biol., pp. 1: 55-59.

Likens G. E., 1975. *Primary Productivity of Inland Aquatic Ecosystem*, *In* Leith H, and Whittaker R.H (eds.), Primary Productivity of the Biosphere, Springer-Verlag, New York, pp:185-202.

Mahajan A., and Kanhere R.R., 1996. *Primary Productivity Studies in the Three Fresh Water Ponds of West Nimar* (M.P), Poll.Res, pp.15 (2):133-135.

Mitra A., Trivedi S., and Choudhary A., 1994. *Interrelationships Between Gross Primary Productivity and Metal Accumulation in the Hoogly Estuary*, Poll. Res., pp. 13 (4):391-334.

Selvi S.J., 2001. *Some Studies on the Environmental Status of Kaliveli Lake and Its Catchment*, MPhil Thesis, Pondicherry Univeristy, p. 147.

Sharma, R., 1993. *Limnological Studies on Yeshwanth Sagar Reservoir with Special Reference to Plankton Population Dynamics*, PhD. Thesis, Devi Ahilya Vishvavidhyala, Indore (M.P).

Shaw B.P., Sahu A., Chowdhury S.B., and Panigrahi A.K., 1991. *Mercury in the Rushikulya River Estuary*, Mar. Polln. Biol, pp. 19: 233-234.

Srinivasan A., 1976. *Limnological Studies and Primary Production in Temple Pond Ecosystems*, Hydrobiologia, pp. 48:117-123.

Vijayaraghavan S., 1971. *Seasonal Variation in Primary Productivity in the Three Tropical Ponds*, Hydrobiologia, pp. 38: 395-408.

Williams A.S., 1972. *Plankton Primary Production in a Tropical Mangrove Bay*, Ophelia, pp. 18: 53-60.

9

Macro Flora and the Plankton of Kaliveli

Abstract

In this chapter, a study on the macro flora and the plankton community has been presented. The study reveals that the macro flora of Kaliveli has been much altered by the local demographic pressures. The flora comprises of 178 species belonging to 60 families representing several rare habit and habitats. Also, the flora is economically significant for its value as food, medicine, thatching/building, ornamental and other purposes.

Kaliveli also supports India's third largest salt marshes, located near the Yedayanthitu estuary.

The studies on the plankton community indicate the lake is dominated by the members of chlorophyceae and bacillariophyceae.

Introduction

Kaliveli supports a wide variety of sedges and grasses which are interspersed with barren sandy areas and muddy margins. When the lake receives adequate rainfall (during the Northeast monsoon, October and November) numerous aquatic plants begin to germinate and grow. The Yedayanthitu estuary, located towards Northeast of the Kaliveli lake, has an abundance of halophytes.

Plate 9.1: The seat of the Lord Ayyanar-the protector of forests and land-at Kaliveli. Sadly, even as some of the inhabitants have invoked the divine blessing, some other inhabitants have ravaged the catchment.

History reveals that during the early 1970's, Kaliveli supported a large and lush mangrove forest (Abbasi, 1997). These forests had been gradually cleared to make way for the agricultural land. At present only a few mangrove bushes remain as relics of the glorious past (Plate 9.1).

In the present chapter, a study on the macro flora and plankton of Kaliveli has been described.

Methodology

Macro flora

The flora of Kaliveli and its surroundings had been collected and identified with the help of *'Flora of the Presidency of Madras'* (Gamble and Fischer, 1935) and *'Flora of Tamil Nadu'* (Nair and Henry, 1983; Henry and Kumari, 1987; Henry et.al., 1989). Before embarking on the survey of flora, *'The Gazetteer of Pondicherry'* has also been consulted.

Plankton

Sample collection. Plankton samples were collected from the subsurface of lake at three sites: Kunimedu and Anumandaikuppam (fresh water part of Kaliveli) and Erapannai (estuarine part of Kaliveli) during the months of November and October 2001.

Concentration of Plankton. The samples were treated with 10ml of liquid detergent and allowed to settle in an lmhoff cone at the rate of 0.5hrs per mm (APHA, 1998; Abbasi, 1998). Later, the supernatant was siphoned off using a suction rubber tube, one end of which was covered with bolting silk cloth (No. 25) as shown in the Fig. 9.1. This was done to avoid the loss of plankton from the supernatant.

The Samples were preserved with 5% formaldehyde and stored at 4° C in polyethylene bottles for further analysis.

Kaliveli being very shallow, often not exceeding 1.5m in depth, the water of the lake is often muddy. The high concentrations of the sediments and decaying plant debris posed difficulties in distinguishing, identifying and quantification of the plankton. Thus, in this chapter we only attempt to present the qualitative studies of the plankton.

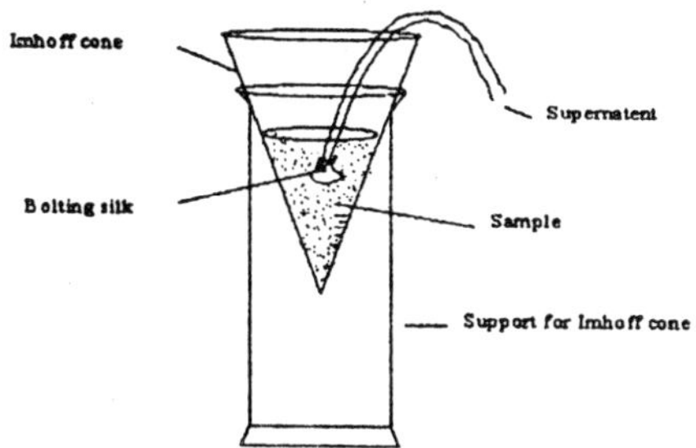

Fig. 9.1: A simple and an efficient device used for concentrating plankton

Results and Discussion

Flora

Kaliveli watershed supports nearly 178 plant species represented by 60 families. Of these, 26 species belong to trees, 56 herbs, 24 shrubs, and 10 climbers (Fig. 9.2). The maximum number of trees was represented by the family euphorbiaceae; and the maximum number of herbs was represented by leguminacae. A complete checklist of all plants in the catchment and waterbody is presented as Table 9.1.

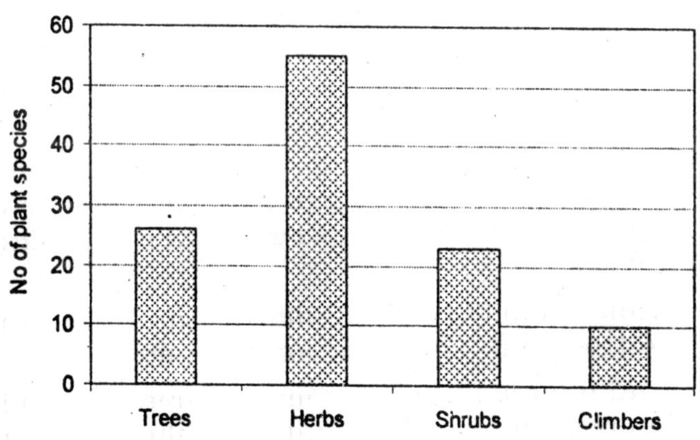

Fig. 9.2: The habit of various plants identified in and around Kaliveli

Table—9.1: Check list of the flora of Kaliveli and its surroundings

Sl. No.	*Family*	*Name of the Species*
1	2	3
1	Acanthaceae	*Adathoda zeylanica*
2	Acanthaceae	*Andrographis paniculata, Nees,*
3	Acanthaceae	*Asystasia gangetica T.And.*
4	Acanthaceae	*Lepidanthus cristata Willd*
5	Aizoaceae	*Corbichonia decumbens*
6	Aizoaceae	*Mollugo cerviana ser;*
7	Aizoaceae	*Mollugo distycha ser*
8	Aizoaceae	*Mollugo pentaphylla Linn*
9	Alangiaceae	*Alangium salvifolium, Wany*
10	Amaranthaceae	*Gomphrena celosioides*
11	Amaranthaceae	*Achyranthus aspera,Linn*
12	Amarathaceae	*Amaranthus spinosus,Linn*
13	Amarathaceae	*Amaranthus virdis, Linn*
14	Amarathaceae	*Digera muricata*
15	Amarathaceae	*Digitana sp.*
16	Amarathaceae	*Celosia polygonoides, Relz.*
17	Anacardiaceae	*Anacardium occidentale, Linn*
18	Apocyanaceae	*Nerium oleander*
19	Apocyanaceae	*Carissa spinarum, Linn.*
20	Arecaceae	*Cocos nucifera*
21	Arecaceae	*Phoenix loureirii*
22	Arecaceae	*Phoenix sylvestris*
23	Aristolochiaceae	*Aristolochia brachteata, Retz;*
24	Aristolochiaceae	*Aristolochia indica, Linn*
25	Asclepiadaceae	*Calotropis gigantia R.Br*
26	Asclepiadaceae	*Caralluma adscendens R.Br.*
27	Asclepiadaceae	*Ceropegia elegans Wall*
28	Asclepiadaceae	*Hemidesmus indicus, R.Nr.*
29	Asclepiadaceae	*Ichnocarpus frutescens R.Br*
30	Asclepiadaceae	*Sarcostemma intermedium Done*
31	Asclepiadaceae	*Tylophora indica RBr*
32	Asclepiadaceae	*Wattakaka volubilis*
33	Avicenniaceae	*Avicennia marina*

(Contd...)

1	2	3
34	Boraginaceae	*Coldenia procumbens, Linn.*
35	Boraginaceae	*Heliotropium indicum Linn*
36	Cactaceae	*Opuntia monocantha Haw*
37	Caparidaceae	*Capparis stylosa,D.C*
38	Capparidaceae	*Cadaba frusticosa*
39	Capparidaceae	*Cephalandra indica*
40	Capparidaceae	*Cleome viscosa, Linn.*
41	Ceratophylleae	*Cerotophyllum demiscum*
42	Charaeceae	*Chara sp.*
43	Chenapodiaceae	*Aerva lanata,Juss;*
44	Chenopodiaceae	*Salicornia sp.*
45	Chenopodiaceae	*Suaedia maritima Dumort*
46	Combretaceae	*Combretum albidum*
47	Commelinaceae	*Commelina clavata Clark comm.*
48	Commelinaceae	*Commelina ensifolia R.Br*
49	Compositae	*Eclipta prostrata*
50	Compositae	*Elephantopus scaber Linn*
51	Compositae	*Parthenium hysterophorus*
52	Compositae	*Tridax procumbens L.,*
53	Compositae	*Vernonia cinerea Less.*
54	Convolvulaceae	*Cissampelos Pareira, Linn*
55	Convolvulaceae	*Evolvulus alsionoides Linn*
56	Convolvulaceae	*Ipomoea carnea Jaca*
57	Convolvulaceae	*Ipomoea prescaprae*
58	Cucurbitaceae	*Citrullus colocynthis, Schrad.*
59	Cucurbitaceae	*Cucumis sativus Linn*
60	Cucurbitaceae	*Luffa cylindrica*
61	Cucurbitaceae	*Mukia maderaspatana*
62	Cyperaceae	*Bulbostylis barbata,Kunth*
63	Cyperaceae	*Ceratophyllum demersum*
64	Cyperaceae	*Cyperus articulatus Linn*
65	Cyperaceae	*Fimbristylis aggregata*
66	Cyperaceae	*Kyllinga bulbosa*
67	Cyperaceae	*Kyllinga nemoralisRotb*

(Contd...)

1	2	3
68	Cyperaceae	*Scirpus lenstrus*
69	Euphorbiaceae	*Microcacca mercurialis Benth*
70	Euphorbiaceae	*Acalypha indica, Linn*
71	Euphorbiaceae	*Bridelia retusa, Spreng*
72	Euphorbiaceae	*Croton bonplandianus*
73	Euphorbiaceae	*Euphorbia antiquorum Linn*
74	Euphorbiaceae	*Euphorbia hirta Linn*
75	Euphorbiaceae	*Flueggea leucopyrus Willd*
76	Euphorbiaceae	*Jatropha curcus Linn*
77	Euphorbiaceae	*Jatropha glandulifera Rotb*
78	Euphorbiaceae	*Jatropha gossypifolia Linn*
79	Euphorbiaceae	*Macaranga peltata*
80	Euphorbiaceae	*Phyllanthus amarus Linn*
81	Euphorbiaceae	*Phyllanthus pinnatus Linn*
82	Euphorbiaceae	*Riccinus communis Linn*
83	Euphorbiaceae	*Euphorbia thirucalli Linn*
84	Fabaceae	*Cicer arietinum*
85	Fabaceae	*Vigna unguiculata*
86	Fabaceae	*Crotolaria pallida*
87	Fabaceae	*Arachis hypogaea*
88	Fabaceae	*Pongamia Pinnata Vent*
89	Flacourtiaceae	*Flacourtia indica*
90	Graminae	*Arundo donax Linn.*
91	Graminae	*Brachiaria ramosa*
92	Graminae	*Cynodon dactylon Pers*
93	Graminae	*Dactyloctenium aegiptium Beaun*
94	Graminae	*Eleusina coracana*
95	Graminae	*Eragrostis hoschst*
96	Graminae	*Eragrostis ciliaris Linn*
97	Graminae	*Eragrostis riparia Nees*
98	Graminae	*Perotis indica O.ktz*
99	Graminae	*Vetivera zizanoioides*
100	Graminae	*Saccharrum officinarum*
101	Guttiferacae	*Garcinia spicata*

(Contd...)

1	2	3
102	Guttiferacae	*Calophyllum inophyllum Linn*
103	Hernandaceae	*Cassytha filiformis, Linn.*
104	Hydrocharidaceae	*Hydrilla verticillata Royl.*
105	Hydrocharidaceae	*Vallisneria natuns Linn*
106	Labiatae	*Anisomeles indica,Okze.*
107	Labiatae	*Leonotis R.Br*
108	Labiatae	*Leucas aspera Spreng*
109	Labiatae	*Ocimum basilicum.Linn*
110	Labiatae	*Ocimum tenuifolium*
111	Leguminosae	*Butea monosperma*
112	Leguminosae	*Cassia auriculata, Linn.*
113	Leguminosae	*Cassia fistula, Linn.*
114	Leguminosae	*Indigofera tinctoria Linn*
115	Leguminosae	*Abrus precatorius.L.*
116	Leguminosae	*Aeschynomene indica,Linn*
117	Leguminosae	*Albizzia amara, Bervin;*
118	Leguminosae	*Tamarindus indica Linn*
119	Leguminosae	*Thephrosia purpurea Pers*
120	Lianaceae	*Hugonia mystax*
121	Liliaceae	*Asparagus racemosus,wild,*
122	Liliaceae	*Aloe vera,Linn*
123	Liliaceae	*Gloriosa superba Linn*
124	Loranthaceae	*Dendrophthoe falcata*
125	Malvaceae	*Abutilon indicum. G.Don;*
126	Malvaceae	*Bombax ceiba*
127	Malvaceae	*Pavonia zeylanica cav*
128	Malvaceae	*Sida acuta Burn*
129	Malvaceae	*Sida cordifolia Linn*
130	Malvaceae	*Sida rhombifolia Linn*
131	Malvaceae	*Thespesia populnea Cav*
132	Melastomaceae	*Memycelon umbellatum Burnt*
133	Meliaceae	*Azadirachta indica A.Juss*
134	Mimosoidae	*Accacia nilotica*
135	Mimosoidae	*Mimosa Linn*

(Contd...)

1	2	3
136	Mimosoidae	*Prosopis juliflora.*
137	Moraceae	*Ficus benghalensisLinn*
138	Moraceae	*Ficus hispida Linn*
139	Morigaceae	*Moringa oleifera Lamk*
140	Myctaginaceae	*Boerhavia diffusa Linn*
141	Myrtaceae	*Syzygium cumini Gaertn*
142	Myrtaceae	*Eucalyptus globulus Labill*
143	Myrtaceae	*Eugenia bracteata Roxb*
144	Najadaceae	*Najas indica cham*
145	Nelumbonaceae	*Nelombo nucifera*
146	Nymphaceae	*Nympluaea pubescens willd*
147	Oleaceae	*Jasminum angustifolium Vnnl*
148	Oleaceae	*Olea Linn*
149	Orchidiaceae	*Peristrophe paniculata Blume*
150	Palmae	*Borassus flabellifer, Linn*
151	Pandanaceae	*Pandanas canarus warb*
152	Pedaliaceae	*Pedalium murex Linn*
153	Poaceae	*Aristida mutabilis*
154	Poaceae	*Chloris inflata*
155	Poaceae	*Oryza sativa*
156	Portulacaceae	*Portulaca oleraceae Linn*
157	Rhamnaceae	*Zizyphus Xylopyrus Willd*
158	Rubiaceae	*Borreria hispida K.Sch*
159	Rubiaceae	*Canthium coromandelicum*
160	Rubiaceae	*Chloroxylon sweitenia, DC.*
161	Rubiaceae	*Morinda coreia Lin*
162	Rubiaceae	*Oldenlandia corymbosa Linn*
163	Rubiaceae	*Oldenlandia umbellata Linn*
164	Rubiaceae	*Randia dumetorum Lank*
165	Rutaceae	*Glycosmis pentaphylla Correa*
166	Sapindaceae	*Cardiospermum halicacabum Linn*
167	Simroubaceae	*Ailanthus excelsa*
168	Solanaceae	*Datura stramorium Linn*
169	Solanaceae	*Phyasalis minima Linn*

(Contd...)

1	2	3
170	Solanaceae	*Solanum amplexicaulis Linn*
171	Solanaceae	*Solanum surranttense*
172	Verbenaceae	*Gmelina asiatica*
173	Verbenaceae	*Lantana Linn*
174	Verbenaceae	*Vitex negundo Linn*
175	Verbenaceae	*Premna corymbosa R & willd*
176	Verbenaceae	*Stachytarpheta jamaicensis Trimen*
177	Vitiaceae	*Cissus quadrangularis Linn*
178	Zygophyllaceae	*Tribulus terrestris Linn*

The flora represents a wide range of habits and habitats. Species such as *Calotropis gijantia, Cissus quadrangularis, and Lepidanthus cristata,* etc., represent the dry terrains. While the species such as *Coldenia procumbens*—the prostrate herb, *Arundo dondax, and Stachytarpheta jamaicensis* etc. represent the wetland types.

Even species representing rare habitats were found such as *Ficus hispida*—a unique representative of the evergreen forests, *Mukia madaraspetania*—this climber belonging to the family of Cucurbitaceae is found usually in the Western Ghats and Nilgiris; *Zizyphus xylopyrus* the dry deciduous forest species, and *Suaeda maritima,* which flourish near the salt pans; *Hugonia mystax,* the climbing shrub of dry forests; *Ficus hispida*—of evergreen, generally, damp places; and *Cassia auriculata .L,*—the leguminous shrub , leaves of which are used for tanning , and is also an indicator of black cotton soils (Fig. 9.3).

The most dominant of the emergent plants in Kaliveli is *Scirpus sp.*—the cyperaceae member belonging to monocotyledonae. It is seen occupying vast stretches of the fresh water part of the lake after the usual monsoons. This emergent species is widely used for thatching the huts, and is extremely popular as it keeps the huts very cool.

A majority of the species-45%—identified at Kaliveli are useful as food. Twenty one per cent are useful as medicines, 14% for thatching and other building purposes, 10% for ornamental and 7% as fuel (Fig. 9.4).

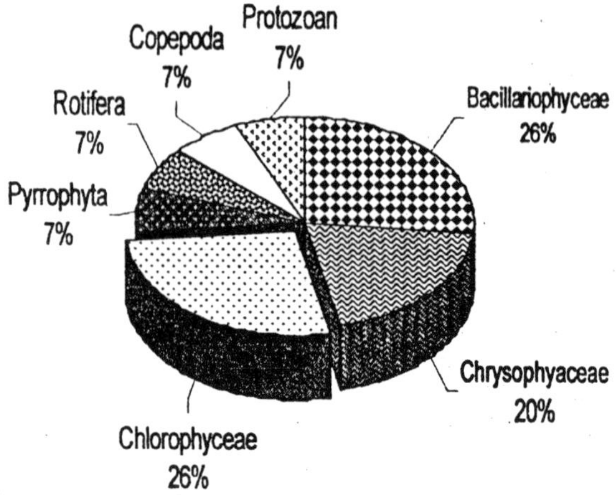

Fig. 9.3: The habitat preference of the various plant species identified in and around Kaliveli.

The ornamental plants include: *Alangium salvifolium* and *Calophyllum inophyllum*

Among the plants of medicinal importance the following are the most common: *Andrographis paniculata, Anisomeles indica, Jatropa curcus, Tamarindus indica, Azadirachta indica,* and *Vitex nedungo.*

Due to the intense demographic pressure and the human interference much of the natural vegetation of Kaliveli and its surroundings has been altered to pave way for the growing demand of agriculture.

The main crop grown is paddy *(Oryza sp.)*, and the subsidiary crops grown are sugar cane *(Saccharrum officinarum)*, ground nut *(Arachis hypogaea)*, *(Eleusine coracana)*, coconut *(Cocos nucifera)*, cow pea *(Vigna unguiculata)*, and bengal gram *(Cicer arietinum)*.

Aquatic Weeds-an Omnius Trend

Kaliveli is being increasingly menaced by aquatic weeds. The lake water is choked with one of the world's worst submerged aquatic weeds: *Ceratophyllum demersum* and *Chara sp*. One can see the blooms of *Chara sp.*, which are being continuously washed ashore by the waves generated in the lake on windy days.

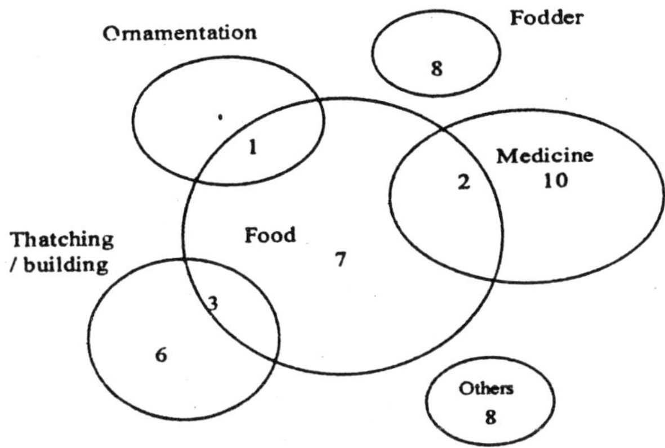

Fig. 9.4: The economic significance of the plant species identified at Kaliveli; the numbers in the figure represent the number of plant species

The continuous input of nutrients from external sources may upset the fragile nutritional balance of the lake. If the lake is subject to further pollution, it may be supplied, by more nutrients from external sources than is needed for its ecological balance, which may encourage the growth of aquatic weeds.

The presence of rampaging mats of aquatic weeds in Kaliveli thus indicate that the lake is being polluted and is, as a result, becoming eutrophic or 'obese' (Abbasi, 1997).

Colonization of aquatic weeds can upset the ecosystem in several ways:

(i) The thick mats of the weeds prevent sunlight from reaching the submerged flora and fauna, thereby cutting off their energy sources and disturbing the ecosystem.

(ii) Once weeds colonise a water body with the help of pollution, then generate their own pollution (Abbasi, Nipaney, 1993),.

The decay of weeds adds to the depletion of dissolved oxygen and increases the BOD, COD, nitrogen, phosphorus, and pathogens.

(iii) The spread of weeds in the lake reduces the area available to fishes and hinders their mobility.

The depletion of dissolved oxygen may result in fish kills or may favour only certain kinds of fishes that can tolerate low oxygen levels.

(*iv*) The profuse growth of weeds breaks natural water currents. The water becomes stagnant favouring the conditions for the breeding of mosquitoes and other disease vectors.

(*v*) Aquatic weeds are known to carry plant pathogens which infect several crops.

The water extracts of fresh and decaying leaves and rhizomes of aquatic weeds are known for phytotoxicity (Ahmed et. al., 1982).

(*vi*) The weeds provide ideal habitat for the growth of mollusks which in turn choke water supply and impart undesirable taste and odour to water.

Also, the snails play a crucial role in the life cycle of blood and liver flukes, shelter, multiply, and find sustenance among the weed roots.

Salt Marshes

The salt marshes at Thenpakkam, in the northeastern portion of Kaliveli, is among the three largest salt marshes of India. The salt marsh is ~3 Km long and 0.5 Km wide. Its vegetation is composed of typical halophytes, occuring as small patches here and there; the prominent species being *Avicennia marina*.

The *Suaeda sp*. is found along the bunds of salt pans. The glassworts—*Salicornia sp*—is seen growing along the inter-tidal zone.

Plankton

The plankton community of Kaliveli was dominated by chlorophyceae and baccillariophyceae (26 %, 4 species each), followed closely by chrysophyceae (20% of the species identified, 3 species each). The rest of the plankton comprised of pyrrophyta, rotifera, copepoda, and protozoa, with single species each (7% of the species identified) (Table 9.2, Fig. 9.5).

Table—9.2: Plankton of Kaliveli during the study

Class	*Species*
Phytoplankton	
Bacillariophyceae	*Asterionella sp.*
	Navicula sp.
	Stephanodiscus sp.
	Surirella sp.
Chrysophyaceae	*Biddulphia sp.*
	Coscinodiscus sp.
	Rhizosolenia sp.
Chlorophyceae	*Chlorella sp.*
	Mougeotia sp.
	Chlamydomonas sp.
	Chaetoceros sp.
Pyrrophyta	*Ceratium sp.*
Zooplankton	
Rotifera	*Daphnia sp.*
Copepoda	*Cyclops sp.*
Protozoan	*Paramoecium sp.*

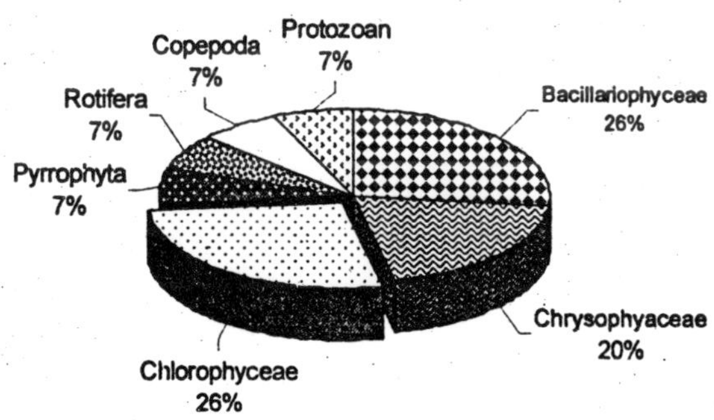

Fig. 9.5: The distribution of various plankton in Kaliveli during the study.

Among the plankton identified at Kaliveli, *Chaetocerous sp.* is known as estuarine pollution alga (APHA, 1998), *Ceratium sp.* which belongs to pyrrophyta, and *Asterionella sp.* belonging to bacillariophyceae, cause unpleasant taste and odour in the water in which they are found. Also, the presence of *Chlamydomonas sp.* indicates fresh water pollution.

REFERENCES

Abbasi and Nipaney 1993 *Worlds Worst Weed— Impact and Utilization*, Industrial Book Distributors, Dehradun, p: 227.

Abbasi S.A., 1997. *Wetlands of India—Ecology and Threats: the Ecology and the Exploitation of Typical South Indian Wetlands*, Vol. I, Discovery Publishing House, New Delhi, p: 149.

Abbasi, S.A., 1998. *Water Quality: Sampling and Analysis*, Discovery Publishing House, New Delhi, (1998), p. 212.

Ahmed, S.A., Ito M, and Neki.K., 1982. *Weed Research*, p: 177.

APHA, 1998. *Standard Methods for the Examination of Water and Waste Water*, American Public Health Association, Washington DC.

Gamble J.S., and Fischer C.E.C., 1935. *Flora of the Presidency of Madras*,3 Vols, Adlard and Son, London, p: 2017.

Henry A.N, Kumari G.R and Chitra V., 1987. *Flora of Tamil Nadu*, Vol 2 Botanical Survey of India, Coimbatore, p: 258.

Henry A.N., Chitra V., and Balakrishnan, N.P., 1989. *Flora of Tamil Nadu*, Vol 3, Botanical Survey of India, Coimbatore, p: 171.

Nair N.C., and Henry A.N, 1983. *Flora of Tamil Nadu*, Vol 1 Botanical Survey of India, Coimbatore, p: 184.

10

Fishes of the Kaliveli Lake

Abstract

In this chapter a study on the fish fauna of Kaliveli has been presented. The assessment was done at two sites: Anumandaikuppam (fresh water part of Kaliveli) and Erapannai (estuarine part of Kaliveli, also called as Yedayanthitu estuary). The species composition and structure of the fish at both the sites has been discussed in view of the present threats that the lake is facing.

Introduction

Fish are good indicators of long-term (several years) anthropogenic impacts on the water quality and the habitat conditions of an aquatic ecosystem. As fish species occupy trophic levels much higher than most other aquatic animals, pollutants get biomagnified in the Piscean tissues. Thus, fish community structure may reflect the health and the recent history of a water body as dramatically as does the presence of aquatic weeds. Moreover, fish are relatively easy to collect and identify. Their environmental requirements, life history information, and distribution are also well known (Plafkin et al. 1989). Above all, fish represent a very important protein source for humans and any adverse impact on the fisheries have serious economic implications.

In this chapter we present studies on the species composition, zoo-mass, and spatial distribution of fishes in the fresh water and the estuarine portions of the Kaliveli wetland. These aspects have been analyzed in the context of the ecological threats to Kaliveli and the strategy for the wetland's conservation.

Materials and Methods

The study was conducted in the month of November, when the lake is usually full due to the northeast monsoon. Fishing net with an eye size of 2x2 cm, with 3 m width and 100 m length was used for catching the fish. The nets were placed in such a fashion that the net spreads vertically from the surface of the lake to the bottom.

One fishing net each was placed for 12 hrs from the evening 7.00 pm till morning 7.00 am at Anumandaikuppam, the fresh water part of the lake, and Erapannai, the estuarine part of the lake (Fig. 10.1).

The captured fish were segregated and weighed within 2 hrs. The representative fish samples were then labelled and preserved in 4% formaldehyde for further identification. The scientific names of the fishes follow Day (1989).

Results and Discussion

A total of 152 fishes, comprising of 20 species and 14 families were picked up in the course of the sampling. It was seen that the estuarine part of Kaliveli (Erapannai) was richer and more diverse in terms of fish fauna compared to the fresh water part (Anumandaikuppam) (Table 10.1, Fig. 10.2). The sampling at Erapannai yielded 144 animals comprising 16 species with a biomass of 4330 g (Table10.1, Fig. 10.2). These species belonged to eleven families.

The most abundant of the species were represented by *Nematolosa nasus*, a total of 86 individuals, weighing 2000 g, amounting to 46.2% of the total biomass of the fishes captured at Erapannai (Fig. 10.3). The species *Chanos chanos*, belonging to the family Chanidae was the next most abundant species at Erapannai, with 34 individuals, weighing 1700 g, amounting to 39.3% of the total fish biomass.

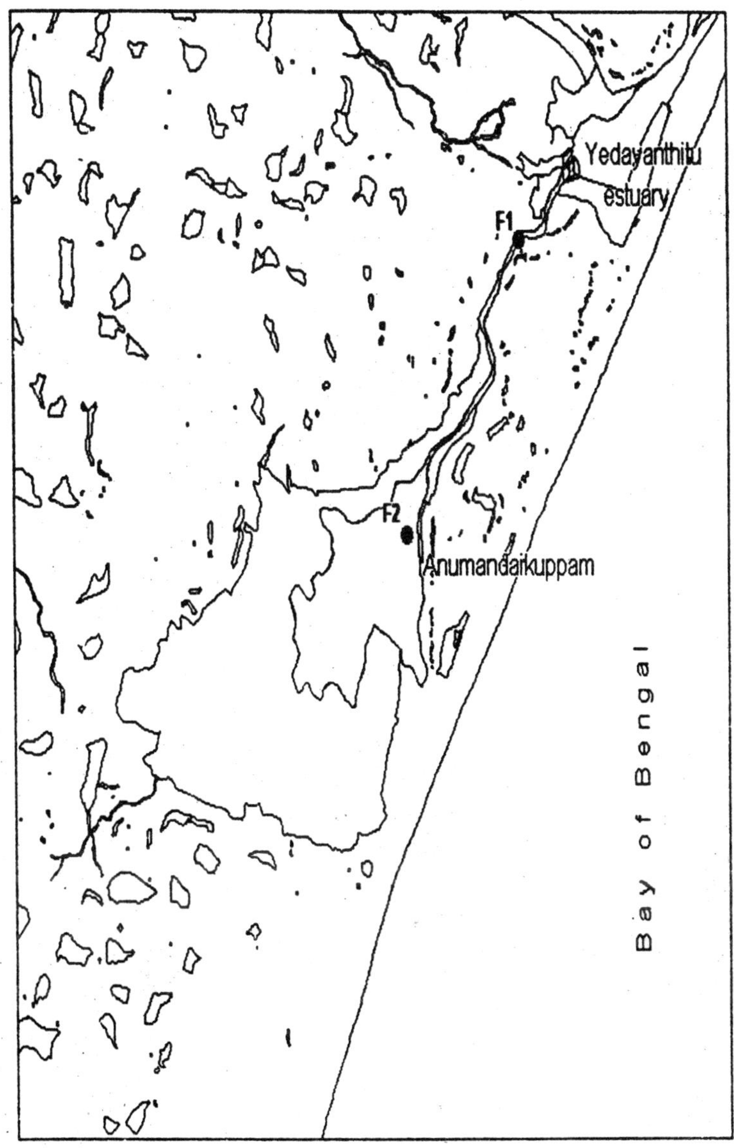

Fig. 10.1: Location of the sampling sites at Kaliveli for the study of fish

Table—10.1: The fish identified at Erappanni, the estuarine part of Kaliveli, and Anumandaikuppam, the fresh water zone of Kaliveli

S.No	*Family*	*Fishes*	*Individuals*	*Weight (g)*	*Biomass (%)*
	Erapani				
1	Clupeidae	Nematolosa nasus	86	2000	46.19
2	Chanidae	Chanos chanos	34	1700	39.26
3	Percidae	Gerres filamentosus	9	150	3.46
4	Mugilidae	Mughil cephalus	2	125	2.89
5	Mugilidae	Mugil cunnesius	6	89	2.06
6	Platycephalidae	Platycephalus indicus	1	50	1.15
7	Gobiidae	Glassogobius giurus	1	50	1.15
8	Percidae	Therapon jarbua	3	40	0.92
9	Unknown	Unknown species1	2	30	0.69
10	Carangidae	Caranx sp.	3	20	0.46
11	Clupeidae	Engraulis mystax	3	20	0.46
12	Labyrinthici	Anabas testudieus	1	20	0.46
13	Siluridae	Macrones seenghala	1	20	0.46
14	Scombrescocidae	Hemiramphus sp.	1	15	0.35
15	Percidae	Ambassis commersoni	1	0.5	0.01
16	Pleuronectidae	Cynoglossus sp.	1	0.5	0.01
	Anumandaikuppam				
17	Chanidae	Channa punctatus	8	220	36.85
18	Gobiidae	Glassogobius giurus	2	220	36.85
19	Siluridae	Wallago attu	4	105	17.59
20	Labyrinthici	Anabas testudieus	4	52	8.71

Among the 16 species of fish identified at Erapannai, 9 species (56.3%, of the total) are known to inhabit both fresh and marine waters in their lifecycle (Fig. 10.4). Young ones of the species, *Ambassis commersoni*, essentially inhabit fresh waters while the adults prefer saline waters (Day, 1989). The species least abundant at Erapannai, *Cynoglossus sp.*, is usually found at the bottom of the sea, and might have ascended the sea and come to the estuary during the high tide.

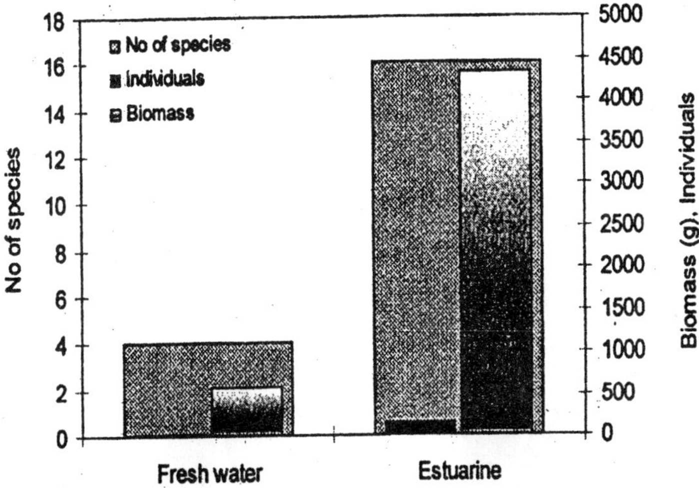

Fig. 10.2: The distribution of fishes in the fresh water part of Kaliveli (Anumandaikuppam) and estuarine part of Kaliveli (Erapannai).

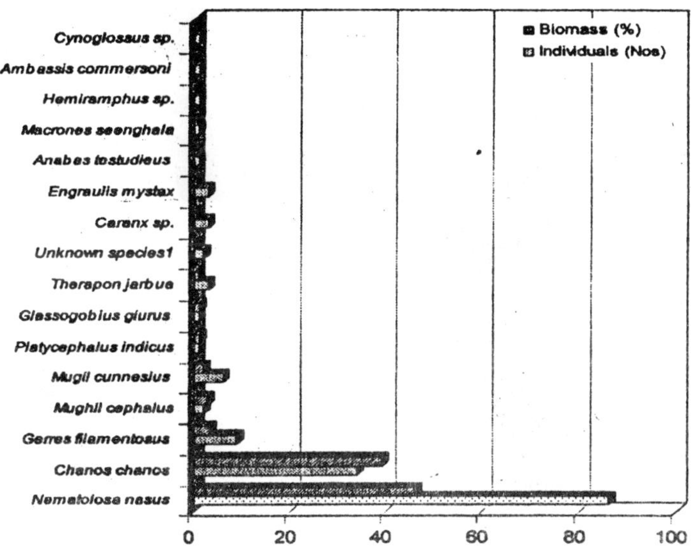

Fig. 10.3: The number of individuals and the relatives biomass (%) of the fishes identified at Erapannai, Kaliveli, during the study

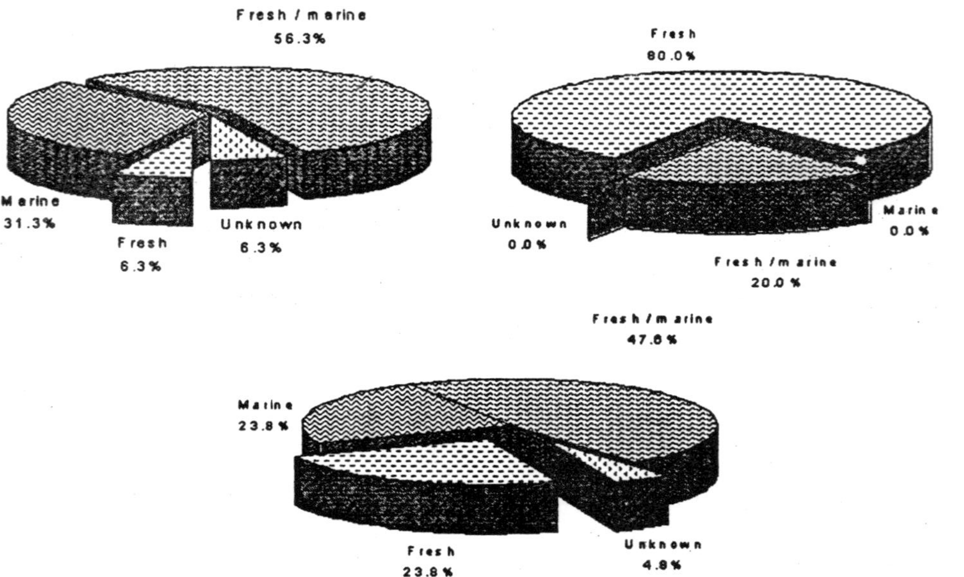

Fig. 10.4: The habitat preference of the fishes identified at (a) Erappani (b) Anumandaikuppam and (c) at both the sites together

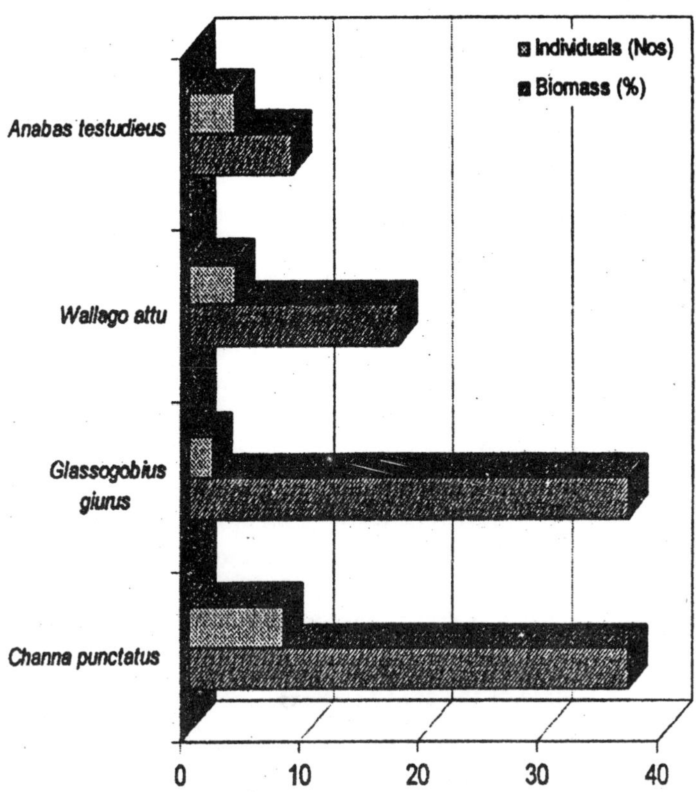

Fig. 10.5: The number of individuals (Nos) and the relative biomass (%) of the fishes identified at Anumandaikuppam, Kaliveli, during the study

At Anumadaikuppam, the fish catch comprised of only four species, each one representing a different family. The most abundant among these were *Channa punctatus*, with 8 individuals weighing 220 g, and *Glassogobius giurus*, with 2 individuals weighing 220 g (Fig. 10.5).

Of the four species identified at Anumandaikppam three were exclusive fresh water inhabitants, while the other species *Anabas testudieus* inhabits both fresh water and estuaries as well. The species *A. testudieus* and *Channa punctatus* can survive for considerable periods out of their usual niche, as these fishes are known to respire directly from air (Day, 1989).

The lesser diversity and the number of fishes in the fresh water part of the Kaliveli could be attributed to the harsh biophysical environment prevalent in this zone. The average depth of the fresh water part of the wetland is not more than a meter, and the water is mostly turbid, making it difficult for the fishes to survive. Also, this part of the wetland dries up very rapidly, in a span of less than 3-4 months. Even during this short span, once the monsoons start retreating (after December), the saline water form the estuary flows into the fresh water zone making the latter saline. Thus, Kaliveli harbours fresh water only for a very short period of about 2 months; the fresh water fish have to complete their life cycle within this period.

Nine of the species found at Erapannai and one of the species found at Anumandaikuppam are known to inhabit fresh water habitats as well as estuaries. However surprisingly none of the fishes identified at Erapannai were found at Anumandaikuppam and vice versa. The only fish caught at both the locations happened to be *Glassogobius giurus*. As the natural habitat of *G. giurus* is fresh water, its occurrence at Erapannai could be due to the straying of this fish into the estuary under the tidal influence.

In the course of the study several fisherman expressed the fear that the shrimp hatchlings and the hatchery effluent which get periodically washed off from the shrimp farms into Kaliveli would cause a decline in the population of commercially important fish. However, the present study reveals that the species composition of native fishes has not been altered much thus far (Chari, 2002). Among the fishes that were captured during the study, prawns *(Paleomon sp.)* comprised of a mere 1.8% of the total fish biomass; and the number of individuals was only 5% when compared with the rest of the species.

Conclusions and Recommendations

The fish composition and community structure showed a very clear distinction between fresh water and estuarine water of Kaliveli. The fish composition was richer and more diverse in the estuary when compared to the fresh water zone. Only one species, *Glassogobius giurus* was found to occur in the fresh water as well as the estuarine portion of Kaliveli.

The recent increase in the shrimp farms along the line of Yedayanthitu estuary does not appear to have contributed to any significant change in the species composition of the native species. The commercially important prawn, *Paleomon sp.*, was rarely encountered and was clearly outnumbered by the native species. However, the rapid conversion of estuary into saltpans and the increasing growth of shrimp farms may lead to the loss of habitat and food for the fishes.

The conversion of vast stretches of the estuary into saltpans would result in the loss of habitat for several fishes. The recent mushrooming of the shrimp farms along the stretch of the wetland that connects Kaliveli lake and Yedayanthithu estuary may alter the water quality due to the release of high nutrient effluents laced with the antibiotics and pesticides that are used in shrimp farming.

The villages at the southern end of the Kaliveli (fresh water zone) is prone to heavy floods during the Northeast monsoons. Recently there have been plans to link the fresh water wetland with sea to prevent these floods. Once the wetland is connected to the sea at the southern end, this would make the present fresh water zone of the wetland brackish and subsequently might alter the community structure and composition of the fishes, eliminating the fresh water fishes.

Vast stretches of the Yedayanthitu estuary has been encroached by saltpans which would directly affect the piscifauna by depriving them of their breeding habitats. Also, the encroachments would drastically reduce the space for the fishes which presently move freely between the sea and the estuary. This would in turn affect the population and diversity of the marine fishes which are known to use the estuaries and the mangroves as the breeding habitats.

It is also imperative that the mangrove forest which existed in the northeastern side of Kaliveli till it was cleared during the 1970's is reestablished. This would go a long way in buffering the impact of storms and providing habitat for the waterfowl, fishes, and other fauna.

REFERENCES

Chari K.B., 2002. *Application of GIS and Remote Sensing in the Environmental Assessment of Two Wetlands of Peninsular India*, PhD Thesis, Pondicherry University, p. 436.

Day F., 1989. *The Fauna of British India FISHES*, Vol. 1 and 2. Today and Tomorrow's Publishers, India.

Plafkin J.L., Barbour M.T., Porter K.D., Gross S.K., Hughes R.M., 1989. Rapid Assessment Protocols for Use in Streams and Rivers: Benthic Macro-invertebrates and Fish, EPA: Washington, D.C.

11

Avifauna of Kaliveli

Abstract

This chapter summarizes the composition, richness, diversity, and population density of the avifauna at Kaliveli. With reference to avifauna, an attempt has been made to establish the link between Kaliveli and Oussudu lake, which is located 15 km south of Kaliveli. Also, the imminent threats that the avifauna of Kaliveli is now facing from have been discussed.

Introduction

Kaliveli, because of its rich diversity of habitats and sheer vastness, attracts a profusion of avifauna. A large number of diverse species of birds on their way to Siberia and other countries halt at Kaliveli every year.

The importance of Kaliveli as a habitat for avifauna has been emphasized by several authors. Perennou (1987), and Perennou and Santharam (1990), who have conducted extensive surveys of waterbirds along the Coromondal coast in 18 major and some smaller wetlands, have described Kaliveli, alongside Pulicat lake, as the most important wintering sites for waterbirds. According to them (Perennou and Santharam 1990), these lakes held altogether 45% of the total duck population including most of the wigeons and red-crested pochards, all the flamingos, large congregations of waders, storks, and egrets. The number of ducks (46,000 and 33,000 in Kaliveli and Pulicat respectively) provide a clue to the international importance of these wetlands.

In this chapter, efforts have been made to review the species composition, richness, diversity, and habitat distribution of avifauna of Kaliveli.

Methodology

The study was made during the period November 1999 and January 2000, when the lake was full after the northeast monsoon. Also, these are the months during which most of the migratory birds visit Indian wetlands to escape the very harsh cold winters of North Asia.

As Kaliveli is a very large wetland, we could only attempt identifying the various birds during the study period. We have also reviewed the work of Krishnan (Scott, 1989), Pieter (Scott, 1989) in the light of our observations.

The diversity of the avifauna was assessed in terms of Simpson's, and Shannon-Wiener's indices. The calculations were assisted by the package MVSP (MultiVariate Statistics Package) marketed by Warren (1993). Locational status and habitat classifications were determined based on the definitions of Ali (1996).

Results and Discussion

Species Composition and Diversity

Kaliveli harbours a rich diversity of birds. So far nearly 183 species belonging to 47 families have been spotted at Kaliveli (Table 11.1). As per the earlier reports of Krishnan (Scott, 1989), Pieter (Scott, 1989) an average 36,671 birds visit Kaliveli during the peak winters. During the year 1988 nearly 81,600 birds visited Kaliveli consisting 15 species.

As shown in the fig. 11.1 and 11.2, the species richness and the total number of birds visiting Kaliveli seem to have decreased drastically. An 85% decrease of the birds was observed between the years 1987 and 1989, and 80% during the subsequent years.

Table—11.1: Avifauna of Kaliveli

S. No	Family	Common name	Scientific Name
1	2	3	4
1	ACCIPITRIDAE	Pariah Kite	*Milvus migrans*
2		Brahminy Kite	*Haliastur indus*
3		Shikra	*Accipiter badius*
4		Whitebellied Sea Eagle	*Haliaeetus leucogaster*
5		White Scavenger Vulture	*Neophron percnopterus*
6		Pale Harrier	*Circus macrourus*
7		Montagu's Harrier	*Circus pygargus*
8		Pied Harrier	*Circus melanoleucos*
9		Marsh Harrier	*Circus aeruginosus*
10		Osprey	*Pandion haliaetus*
11		Black winged kite	*Elanus caeruleus*
12		Black eared kite	*Milvus lineatus*
13		Lessar Grey-headed fishing Eagle	*Ichthyophaga ichthyaetus*
14		Hen Harrier	*Circus cyaneus*
15		Short-toed Eagle	*Circaetus gallicus*
16	ALAUDIDAE	Bush lark	*Mirafra assamica*
17		Sky Lark	*Alauda arvensis*
18		Redwinged Bush Lark	*Mirafra erythroptera*
19		Ashycrowned Finch Lark	*Eremopterix grisea*
20		Rufous Tailed Finch Lark	*Ammomanes phoenicurus*
21		Oriental Skylark/Indian Small Sky Lark	*Alauda gulgula*
22	ALCEDINIDAE	Lesser Pied Kingfisher	*Ceryle rudis*
23		Common Kingfisher	*Alcedo athis*
24		Whitebreasted Kingfisher	*Halcyon smyrnensis*
25	ANATIDAE	Ruddy Shelduck	*Tadorna ferruginea*
26		Pintail	*Anus acuta*
27		Wigeon	*Anas penelope*
28		Garganey	*Anas querquedula*
29		Shoveller	*Anas clypeata*
30		Common Teal	*Anas acuta*

(Contd...)

1	2	3	4
31		Spotbilled Duck	*Anas poecilorhyncha*
32		Redcrested Pochard	*Netta rufina*
33		Common Pochard	*Aythya ferina*
34	*APOBIDAE*	Palm Swift	*Cypsiurus parvus*
35		House swift	*Apus affinis*
36	*ARDEIDAE*	Grey Heron	*Ardea cinerea*
37		Large Egret	*Ardea alba*
38		Paddybird	*Ardeola grayi*
39		Median/Smaller Egret	*Egretta intermedia*
40		Little Egret	*Egrette garzetta*
41		Indian Reef Heron	*Egretta gularis*
42		Purple Heron	*Ardea Purpurea*
43		Pond Heron	*Ardeola grayii*
44		Cattle Egret	*Bubulcus ibis*
45	*ARTAMIDAE*	Ashy Swallow Shrike	*Artamus fuscus*
46	*BURHINIDAE*	Stone Curlew	*Burhnus oedicnemus*
47	*CAMPEPHA-GIDAE*	Common Wood Shrike	*Tephrodornis pondicerianus*
48	*CAPITONIDAE*	Small Green Barbet	*Megalaima viridis*
49		Coppersmith or Crimson-brested Barbet	*Megalaima haemacephala*
50	*CHARADRII-DADE*	Yellow-wattled lapwing	*Vanellus malabaricus*
51		Grey-headed lapwing	*Vanellus leucurus*
52		Ringed plover	*Charadrius dubices*
53		Kentish plover	*Charadrices abxandrinus*
54		Red Wattled Lapwing	*Vanellus indicus*
55		Grey Plover	*Pulvialis squatarola*
56		Eastern Golden Plover	*Pulvialis dominica*
57		Large Sand Plover	*Charadrius leschenaultii*
58		Little Ringed Plover	*Charadrius dubius*
59		Kentish Plover	*Charadrius alexandrinus*
60		Lesser Sand Plover	*Charadrius mongolus*
61		Whimbrel	*Numenius phaeopus*
62		Curlew	*Numenius arquata*
63		Black tailed Godwit	*Limosa limosa*

(Contd...)

1	2	3	4
64		Turnstone	*Arenaria interpres*
65		Pintail Snipe	*Gallinago stenura*
66		Fantail Snipe	*Gallinago gallinago*
67		Long-toed Stint	*Calidris subminuta*
68		Dunlin	*Calidris alpina*
69		Ruff	*Philomachus pugnax*
70	*CICONIIDAE*	Painted Stork	*Mycteria leucocephala*
71		Openbill Stork	*Anastomus oscitans*
72		White Stork	*Ciconia ciconia*
73	*COLOMBIDAE*	Blue Rock Pigeon	*Columba livia*
74		Spotted Dove	*Streptopelia chinensis*
75		Turtle Dove	*Streptopelia turtur*
76		Indian Ringed Dove	*Streptopelia decaocto*
77		Little Brown Dove	*Streptopelia senegalensis*
78	*CORACIIDAE*	Indian Roller	Coracias benghalensis
79	*CORVIDAE*	Indian Treepie	*Dendrocitta vagabunda*
80		House Crow	*Corvus splendens*
81		Jungle Crow	*Corvus macrorhynchos*
82	*CUCULIDAE*	Common Hawk-cuckoo	*Cuculus varius*
83		Pied Crested Cuckoo	*Clamator jacobinus*
84		Koel	*Eudynamys scolopacea*
85		Coucal (or) crow-pheasant	*Centropus sinensis*
86	*DICRURIDAE*	Black Drongo	*Dicrurus adsimilis*
87	*FALCONIDAE*	Peregrine Falcon	*Falco peregrinus*
88		Kestrel	*Falco tinnunculus*
89	*GLAREOLIDAE*	Indian Courser	*Cursorius coromandelicus*
90		Large Indian Pratincole	*Glareola pratincola*
91		Small Indian Pratincole	*Glareola lactea*
92	*HIRUDINIDAE*	Striated or Red-rumped Swallow	*Hirundo daurica*
93		Collared Sandmartin	*Riparia riparia*
94		Swallow	*Hirundo rustica*
95	*IRENIDAE*	Common Iora	*Aegithina tiphia*
96	*LARIDAE*	Great Blackheaded Gull	*Larus ichthyaetus*
97		Brownheaded Gull	*Larus brunnicephalus*

(Contd...)

1	2	3	4
98		Blackheaded Gull	*Larus ridibundus*
99		Whiskered Tern	*Chlidonias hybrida*
100		White-winged Black tern	*Chlidonias leucopterus*
101		Gullbilled Tern	*Gelochelidon nilotica*
102		Caspian Tern	*Hydroprogne caspia*
103		Common Tern	*Sterna hirundo*
104		Little Tern	*Sterna albifrons*
105		Large Crested Tern	*Sterna bergii*
106		Herring Gull	*Larus argentatus*
107		Lesser Black-backed Gull	*Larus fuscus*
108		Slender-billed Gull	*Larus genei*
109		Little Gull	*Larus minutus*
110		Indian River Tern	*Sterna aurantia*
111		Black-caped Tern	*Chlidonias niger*
112		Black bellied Tern	*Sterna acuticuda*
113	*MEROPIDAE*	Bluetailed Bee-eater	*Merops philippinus*
114		Green Bee-eater	*Merops orientalis*
115	*MOTACILLIDAE*	Paddyfield Pipit	*Anthus novaeseelandiae*
116		Richards Pipit	*Anthus richardi*
117		Yellow Wagtail	*Motacilla flava*
118		Large Pied Wagtail	*Motacilla maderaspatensis*
119		Forest wagtail	*Motacilla indica*
120		Thick-billed flower-pecker	*Dicaeum agile*
121		Tickell's Flower-Pecker	*Dicaeum erythrorhynchos*
122	*MUSCICAPIDAE*	Streaked Fantail Warbler	*Cisticola juncidis*
123		Franklin's wren-warbler	*Prinia hodgsonii*
124		Plain wren-warbler	*Prinia subflava*
125		Ashy wren-warbler	*Prinia socialis*
126		Indian Great Reed warbler	*Acrocephalus stentoreus*
127		Blyth's Reed warbler	*Acrocephalus dumetorum*
128	*Subfamily SYLVIINAE*	Tailor Bird	*Orthotomus sutorius*
129		Green Warbler	*Phylloscopus nitidus*
130	*Subfamily TIMALINAE*	Whiteheaded Babbler	*Turdoides affinis*

(Contd....)

1	2	3	4
131	*Subfamily TURDINAE*	Magpie Robin	*Copsychus saularis*
132		Indian Robin	*Saxicoloides fulicata*
133	*NECTARINIDAE*	Purple rumed Sunbird	*Nectarinia zeylonica*
134		Loten's Sunbird	*Nectarinia lotenia*
135		Purple Sunbird	*Nectarinia asiatica*
136	*ORIOLIDAE*	Golden Oriole	*Oriolus oriolus*
137	*PELICANIDAE*	Spottedbilled or Grey Pelican	*Pelecanus philippensis*
138	*PHALACRO-CIDAE*	Cormorant	*Phalacrocorax carbo*
139		Little Cormorant	*Phalacrocroax niger*
140	*PHASSIAN-NIDAE*	Grey Partridge	*Francolinus pondicerianus*
141	*PHOENICOPT-ERIDAE*	Greater Flamingo	*Phoenicopterus roseus*
142		Lesser Flamingo	*Phoenicopterus minor*
143	PICIDAE	Lesser Goldenbacked Woodpecker	*Dinopium benghalense*
144	*PLOCEIDAE*	Baya (Baza)	*Aviceda leuphotes*
145		Common Silverbill	*Lonchuria malabarica*
146	*Subfamily PASSERINAE*	House Sparrow	*Passer domesticus*
147		Yellow throated Sparrow	*Petronia xanthocollis*
148	*PODICIPEDIDAE*	Little Grebe	*Tachybaptus ruficollis*
149	*PSITTACIDAE*	Roseringed Parakeet	*Psittacula krameri*
150		Alexandrine Parakeet	*Psittacula eupatria*
151	*PYCNONO-TIDAE*	Redvented Bulbul	*Pycnonotus cafer*
152		White-browed Bulbul	*Pycnonotus luteolus*
153	*RALLIDAE*	White-breasted waterhen	*Amaurornis phoenicurus*
154		Moorhen	*Gallinula chloropus*
155		Purple Moorhen	*Porphyrio porphyrio*
156		Coot	*Fulica atra*
157	*RECURVIRO-STRIDAE*	Blackwinged Stilt	*Himantopus himantopus*
158		Avocet	*Recurvirostra avosetta*
159	*SCOLOPACIDAE*	Spotted Redshank	*Tringa erythropus*

(Contd. . .)

1	2	3	4
160		Redshank	*Tringa totanus*
161		Marsh Sandpiper	*Tringa stagnatilis*
162		Greenshank	*Tringa nebularia*
163		Green Sandpiper	*Tringa ochropus*
164		Wood Sandpiper	*Tringa glareola*
165		Terek Sandpiper	*Tringa terek*
166		Common Sandpiper	*Tringa hypoleucos*
167		Little Stint	*Calidris minuta*
168		Temminck's Stint	*Calidris temminckii*
169		Curlew Sandpiper	*Calidris testacea*
170		Black tailed Godwit	*Limosa limosa*
171		Bartailed Godwit	*Limosa lapponica*
172		Eurasian curlew	*Numenius arquata*
173		Common snipe	*Gallinago gallinago*
174	*STRIGIDAE*	Spotted Owlet	*Athene brama*
175	*STURNIDAE*	Common Myna	*Acridotheres tristis*
176	*THRESKIORN-ITHIDAE*	White Ibis	*Threskiornis aethiopica*
177		Glossy Ibis	*Pseudibis papillosa*
178		Spoonbill	*Platalea leucorodia*
179	*UPUPIDAE*	Hoopoe	*Upupa epops*
180	*ZOSTEROPIDAE*	White-eye	*Zosterops palpebrosa*

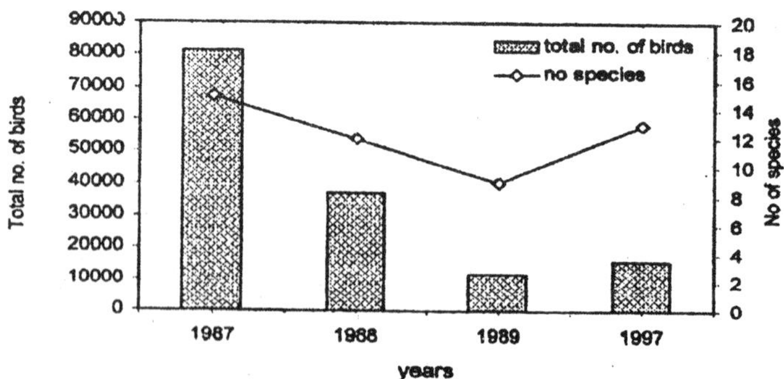

Fig. 11.1: The total number of birds and the number of different species spotted in and around Kaliveli during the years 1987, 1988, 1989 and 1997

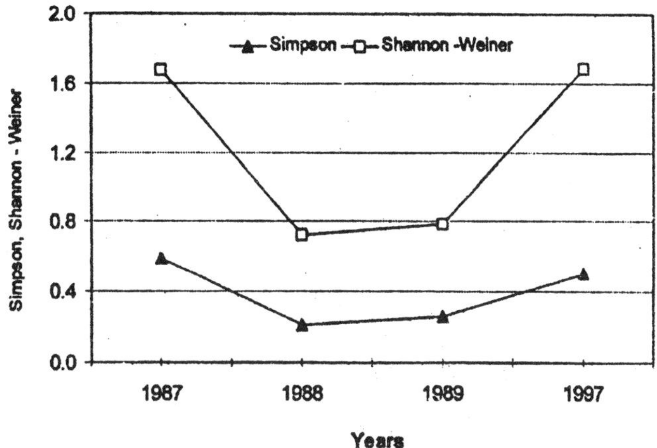

Fig. 11.2: Change in the levels of species diversity indices during the years 1987, 1989, and 1997

The ducks belonging to the family Anatidae with an average abundance of 71% are the most abundant species which is followed by waders (20%). There had been a decrease in the abundance of terns, glossy ibis, ducks, waders and open billed stork; while the abundance of the grey heron, white stork and egrets were on the increase (Fig. 11.3).

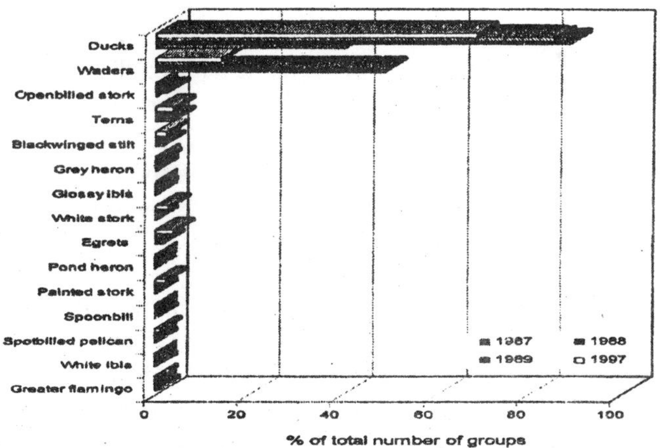

Fig. 11.3: Relative density (%) of avifauna in and around Kaliveli wetland during the years 1987, 1988, 1989, and 1997

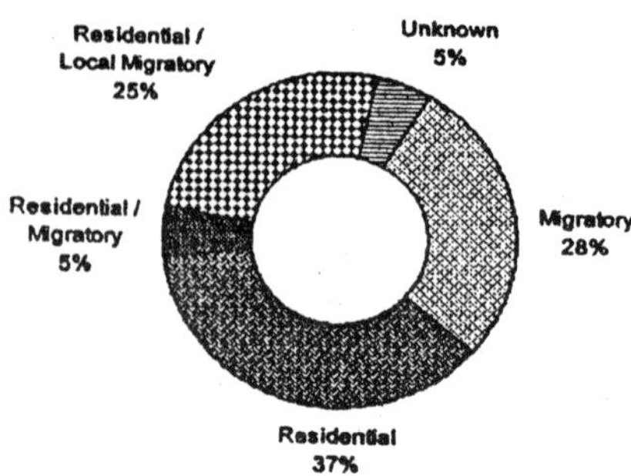

Fig. 11.4: Locational status of the Kaliveli avifauna. The local-migratory refers to the birds that migrate within the geographical limits of India; unknown, refers to the birds that could not be identified during the study

A majority of the bird species identified at Kaliveli are residential, 38% of the total number of species (Fig. 11.4). Nearly 28% species are migratory, 25% residential/local migratory and 5% residential migratory.

The migratory birds were represented by the families Laridae (12 species), Scolopacidae (10 species), Charadriidae (8 species), Anatidae (4 species), Moticillidae, Muscicapidae: 3 species each, Apodidae, Falconidae: 2 species each, and Accipitridae, Alaudidae, Ciconiidae, Colonibidae, Hirudinidae, Floceidae: one species each.

The ecological wealth of Kaliveli is endorsed by the presence of a number of predators-birds of prey. The white-bellied sea-eagle, and spotted eagle, a species which had not been seen in South India for decades are a few of them (Abbasi, 1997). Due to the unfavorably altering ecological conditions, the spotted eagle had probably vanished from the South (Ali and Ripley, 1983).

Breeding Pattern

A close look at the breeding pattern of the various birds at Kaliveli (Table 11.2) indicate that nearly 7.3% (13 species) breed during the months of April to June, 4% breed between March to August and another 4% breed during May to June. Nearly 4% of the total birds, constituting 7 species breed anytime in the months during January till December.

Table—11.2: The breeding pattern of birds (number of species) in the Kaliveli region along the year. (For instance the number of bird species that breed between Jan-June are 2, between Jan-July are 2 and Jan-Nov are 1)

End of the breeding period ↓	*Beginning of the breeding period*												
	Jan	*Jan*	*Feb*	*Mar*	*Apl*	*May*	*Jun*	*Jul*	*Aug*	*Sep*	*Oct*	*Nov*	*Dec*
Jan													
Feb										1			
Mar							1	1				3	
Apl			3	1						1		2	3
May			4	4									1
Jun		2	2	3	13	7					1		1
Jul		2	3	5	5	4	4		1				
Aug			1	7	4		2	3					
Sep			2	4	1	2	4	3					
Oct			1		3		2						
Nov		1						1					
Dec													

Among the migratory birds 3 species breed during the period April-June, four species breed during May-June, and one species breed during the period March-June. Some of the migratory birds which prefer Kaliveli as a roosting site are: white stork *(Ciconia ciconia)*, peregrine falcon *(Falco peregrinus)*, Swallow *(Hirundo rustica)*, brown headed gull *(Larus brunnicephalus)*, forest wagtail (Motacilla indica) and streaked fan tail warbler *(Cisticola juncidis)*.

Habitat Distribution

The diverse habitats found around Kaliveli, such as agriculture fields, marshes, fresh water and estuarine water, seem to attract a variety of birds sharing different habitats.

In Winter, tens of thousands of ducks arrive and begin fulfilling all their requirements from Kaliveli. Dabbling ducks feed at the shallow, open, margins of the lake. The centre of the water body is used during daytime to escape from the human disturbance coming from the shores.

As the winter ends and the smaller ponds and paddy fields surrounding Kaliveli dry-up, the ducks and other birds congregate on Kaliveli in even bigger numbers than in winter. Perennou (1987) found as many as 1400 egrets (little, smaller, and large) fishing together with 1500 black winged stilts and 40,000 waders on their way back to Siberia!

Of the birds identified at Kaliveli, 67 species prefer a single habitat such as agriculture fields, open lands, wetlands and orchids (Fig. 11.5). A majority of the birds, nearly 79 species, share atleast two different habitats such as open lands and shallow water. Another 20 species share atleast three different habitats.

Of the various habitat distributions, the birds that share open lands and wetlands form the vast majority, followed by the birds that favor wetlands.

The Link Between Kaliveli and Oussudu

The studies of Krishnan (Scott, 1989), Pieter (Scott, 1989), Abbasi (1997), and Chari and Abbasi (2002) have pointed towards the possible ecological link between Kaliveli and Oussudu lake, located some 15 km South of Kaliveli. We have made a comparison between the avifaunal structure and composition of Oussudu lake and Kaliveli.

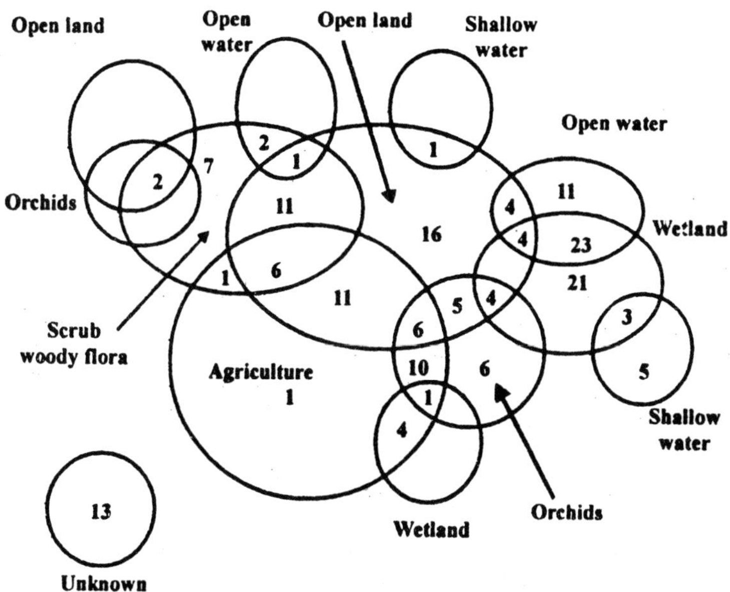

Fig. 11.5: The habitat distribution of various birds found at Kaliveli as per Grimmett *et. al.*

A systematic over 5 years of study was conducted by Chari and Abbasi (2002) on the avifauna of Oussudu lake. The birds identified by these authors have been used for the comparison of avifauna of Kaliveli.

Kaliveli seem to hold a large congregation, and more diverse avifauna than Oussudu. However, Oussudu lake being deeper than Kaliveli, favor birds such as grebes and ducks; and Kaliveli being shallower, favour such birds as storks ibis and egrets. Kaliveli has 90 species which are unique to it, that is the birds found at Kaliveli and not Oussudu; and there are 34 birds which are unique to Oussudu. On the other hand, there are 89 birds that were found at both the lakes (Table 11.3).

Table—11.3: A comparison of the avifauna of Kaliveli and Oussudu lake

		Kaliveli lake	
		No. of species present	*No. of species absent*
Oussudo lake	No. of species present	89	34
	No. of species absent	90	-

The measures of qualitative similarity between Oussudu and Kaliveli are shown in the table 11.4. The Sorenson index, which weighs the positive matches more than it weighs the negative matches, measured 0.596 (Krebs, 1989). The Simple Matching Coefficient, which weighs both the negative and positive matches, measured 0.417, which is less than the Sorenson index (Krebs, 1989). Thus, as per the qualitative similarity indices the avifauna of both the lakes match 50 % approximately.

The birds that are unique to Kaliveli are represented by the families Accipitridae, Charadridade, Laridae, Scolopacidae, Artamidae, Burhinidae, Campephagidae, and Glareolidae. While the birds that are unique to Oussudu are represented by the families Anatidae, Laniidae, Pittidae, Caprimuifisae, and Tytonidae.

The families that represent both Kaliveli and Oussudu are Alcedinidae, Apobidae,

Table—11.4: Qualitative similarity measures of the avifauna of Kaliveli and Oussudu lake

Similarity Measures	*Value*
Coefficient of Jaccard	0.417
Coefficient of Sorenson	0.590
Simple Matching Coefficient	0.417

Threats

The avifauna of Kaliveli is now facing threats from hunting of birds, intensive fishing, and land use conversion (Plate 11.1).

Poaching of birds by several means such as: nets, poisoned fish as bait and shooting is quite common (Chari, 1997). No concrete steps have been taken towards conservation of the wildlife of this region (Plate 11.2).

Plate 11.1: Vast stretches of Yedanthitu estuary has been encroached by shrimp farms (above) and salt pans (below) resulting in the loss of precious habitats for avifauna

Vast stretches of Yedayanthitu estuary is converted into salt pans altering not only the hydrology and water quality but also the entire ecosystem of the region. The agriculture encroachments inside the Kaliveli would alter the habitat and the disturb the roosting and feeding birds.

In the last 3-4 years several shrimp farms have sprang up along the estuarine part of the Kaliveli. These shrimp farms, apart from adding pollution to the lake, may disturb the peacefully perching birds by increasing human activity in the surroundings.

The resources of Kaliveli have been exploited traditionally for a long time. However, in the recent past the intensive fishing, both in Kaliveli and in Yedayanthitu estuary, seem to affect the piscifaunal diversity and thus indirectly it would affect the birds that feed on fishes.

REFERENCES

Abbasi SA. 1997. *Wetlands of India—ecology and Threats: The Ecology and the Exploitation of Typical South Indian Wetlands*, Vol.I, Discovery publications House, New Delhi, p: 149.

Ali S, 1996. *The Book of Indian Birds, Bombay Natural History Society, Oxford University Press*, India, p: 354.

Ali S., and Ripley S.D., 1983. *Handbook of the Birds of of India and Pakistan*, Oxford University Press, p: 625.

Chari K.B., and Abbasi S.A., 2002. *Ecology, Habitat and Community Structure of Avifauna at Oussudu Lake: Towards a Strategy for Conservation and Management*, Aquatic Conservation, Inpress.

Chari KB. 1997. *A Reconnaissance of the Ecology of Two Wetlands : Ousteri and Kaliveli*, MS., Thesis, Pondicherry University, p: 110.

Grimmett R, Carol I, and Tim T, 1998. *Birds of the Indian Subcontinent*, Oxford University Press, Delhi, p: 888.

Krebs CJ. 1989. *Ecological Methodology*, Harper & Row Publishers, New York, p: 425.

Perennou C., 1987. Two Important Wetlands Near Pondicherry, Blackbuck, 3 1-9.

Perennou C., and Santharam V., 1990. *An Anthropological Survey of Some Water Birds*. J. Bombay Nat. Hist. Soc. 87 354-363.

Scott DA. 1989. *A Directory of Asian Wetlands*, IUCN, The World Conservation Union, Gland, Switzerland, and Cambridge, UK., p: 1181.

Warren LK. 1993. *MVSP-a MultiVariate Statistical Package*, Kovach Computing Services, U.K.

SECTION—VI

IN PARTING....

12

In Parting

Overview

In this concluding chapter we present an overview of the contents of the book and catalogue the imperatives that have emerged from the study.

Chapter 1

Introduction

In recent times GIS and remote sensing have emerged as powerful tools in the management of earth resources. In this chapter, we have presented the various definitions, characteristics, components, and the technical elements of a GIS. Also, we have discussed the role played by GIS and remote sensing in wetland management with several illustrative case studies.

Chapter 2

Wetlands—an Overview

Wetlands have always been important but in recent years their value has been increased in direct proportion to the threats wetlands are facing from population and pollution.

This chapter is a state-of-the-art review on the classification, importance, and status of wetlands, with particular emphasis on

Indian wetlands. The role of the two most modern and powerful tools—remote sensing and GIS—in wetland management is highlighted.

Chapter 3

Kaliveli Wetland: A Typical Bulwark Against Drought

Kaliveli is one of the most prominent among wetlands that dot the semi-arid landscape of the peninsular east coast. In this chapter, an overview of the role played by Kaliveli in water storage, groundwater recharge, fisheries, and wildlife protection has been presented. The potential threats of eco-degradation, and the various measures necessary to conserve the wetland have also been discussed.

Kaliveli is now facing severe threat from: (i) poaching of birds (ii) the intensive practice of agriculture in-and-around Kaliveli (iii) dumping of industrial wastes and (iv) drastic shrinking of the water -spread due to encroachments.

The imperatives that require immediate attention for conserving Kaliveli wetland are: (i) declaring Kaliveli a bird sanctuary and designate it as a biosphere reserve under the MAB (man and Biosphere) UNESCO program (ii) conducting a detailed study on the vegetation structure and dynamics of the lake (iii) reforestation of mangrove vegetation and other vegetation in-and-around Kaliveli (iii) developing a programme for creating mass awareness especially among the people living around kaliveli.

Chapter 4

Assessing Land Use/land Cover of Kaliveli Watershed Using Remote Sensing and GIS: Implications for Conservation

In this chapter several features of the Kaliveli wetland were studied by synthesizing the inputs from satellite imagery, topo-sheets, and extensive on-ground surveys. A geographical information system (GIS) was developed for the wetland with the aid of computer-automated tool MapInfo Professional5.5. With the aid of GIS the implications of the present land use on the lake and its water system have been identified. Various measures for conservation of the lake and its watershed have been suggested based on the above findings.

The following conclusions can be drawn from the above study:

(1) If we had assessed the situation purely on the basis of the imagery, little difference would have been discerned between 1970 and 1997. But the ground truth studies reveal that the situation has changed drastically. The items of information not revealed by the imageries but confirmed by the ground survey are:

(a) encroachment and practice of agriculture in and around Kaliveli

(b) mushrooming of shrimp farms near Erapannai and Mudaliarpet.

These trends if not checked immediately can prove detrimental to the lake on the long run.

(2) High and low saline areas occur in the vicinity of the lake due to waterlogged conditions, practice of intensive agriculture, and dry/wet deposition of salt because of the proximity of the sea.

(3) The lake and its potential water spread area need to be earmarked to check further encroachment.

(4) Considering the ecological significance, as substantiated by the present and the earlier studies, Kaliveli should be declared a biosphere reserve.

Chapter 5

Soil Quality of Kaliveli and Its Surroundings

This chapter deals with the soil quality of Kaliveli region, studied vis-a vis the soil texture, pH, electrical conductivity and the nutrient status. Clay was the predominant soil type found in the study area. As clay is known to bind positively charged cations, contribution of the catchment in terms of the soil cations—Ca^{2+}, Mg^{2+}, K^{+}, Na^{+}, and H^{+}, is likely to be negligible. On the other hand, the negatively charged cations, such as Cl^{-1} and No^{3-} , which are not bound by this soil, are likely to be abundant in the run-off.

A majority of the soil samples of Kaliveli region were of neutral pH regime, The acidic soil samples identified in the study area had been attributed to the soil acidification influenced by the continual use of inorganic fertilizers on agricultural soils.

The saline soils identified at Kaliveli can be related to: (i) the over extended period of evaporation and evapo-transpiration of soil, exceeding the downward percolation of rainfall, irrigation water, (ii) 'salt spray' due to the proximity of sea and (iii) 'efflorescence'.

In general, the phosphorus levels appear to be negligible when compared to the nitrogen in the soils found in this region. The relatively high total nitrogen content in some of the soils could be due to the extensive agriculture practice and mineralization of organic substances in those areas.

Chapter 6

Water Quality of Kaliveli Lake

The aim of this chapter is to elucidate the interrelationships of various water quality variables, identify the pollutants, if any, and to determine the trophic status of the lake. The dynamics of lake water quality were discussed over time and space. Where possible, efforts were made to identify the likely source of these pollutants. Also, the water quality of Kaliveli lake was assessed for the criteria of BIS (1994) for fish and aquaculture.

Some of the salient findings of the study were:

(1) Kaliveli is a very shallow and seasonal wetland; the fresh water part of the lake holds water for 3-4 months only. Thus, water quality of the lake changes as rapidly as the quantity of the water over this period.

(2) The chief factors that control the dynamics of Kaliveli water quality are:

(i) direct precipitation (ii) non-point source pollutants—both agriculture and storm run-off (iii) evaporation and (iv) proportionate admixing of the sea water through the estuary.

(3) The Kaliveli watershed being rich in iron, contributes to the lake through runoff. The lake, based on the criteria of total phosphorus had been classified as hyper-eutrophic.

Chapter 7

Ground Water Quality of Kaliveli Lake and Its Surroundings

A detailed study on the ground water quality of Kaliveli and its surroundings has been presented in this chapter. The ground water quality was monitored periodically during the years 2000 and 2001. Water quality parameters such as pH, EC and other 13 essential parameters were analyzed and the results were assessed for the fitness of drinking and irrigation water quality criteria of BIS. The areas with poor ground water quality had been identified using the spatial modelling of GIS.

The present study on the ground water of Kali''eli reveals that precipitation, and the subsequent leaching of the soil constituents and evapo-transpiration are the primary factors that determine the ground water quality. The predominant chemical species that enter the ground water aquifers are nitrogen, phosphorus—where the soils are acidic, and the areas with intensive agriculture practice, and iron—at the areas where water is logged for long time, or , the ground water HRT had been for considerably long.

Kaliveli faces a serious threat of eutrophication because of the following factors: (i) land use/land cover pattern as already mentioned in the chapter 4, (ii) the soil type, which is predominantly clay, can leach the nitrites very readily—both into the ground water aquifers and the lake as run-off, and (iii) the heavy drafting of the ground water for irrigation.

The imminent threat that Kaliveli region is now facing is from the intensive practice of shrimp farming, which would not only aggravate the soil chemistry but also pollutes the ground and surface water.

In recent times there has been a rapid increase in urbanization and industrialization of Kaliveli region. Already, a pharmaceutical industry, an alkali manufacturing unit, and Pondicherry University is located in this watershed. A recent addition to this cluster is a medical college and hospital, with the accessories of doctor's colony. The intensity of urbanization is so rapid that before one can realize, it would be too late to find the remedies that would plague the ground water of this region very soon. In this context it

(1) As mentioned in the chapter 3, the avifauna of Kaliveli is now facing threats from hunting of birds, intensive fishing, and land use conversion.

(2) Poaching of birds by several means such as: nets, poisoned fish as bait and shooting is quite common. No concrete steps have been taken towards conservation of the wildlife of this region.

(3) Vast stretches of Yedayanthitu estuary has been converted into salt pans altering not only the hydrology and water quality but also the entire ecosystem of the region. The agriculture encroachments inside the Kaliveli would alter the habitat and disturb the roosting and feeding birds.

(4) In the last 3-4 years several shrimp farms have sprang up along the estuarine part of the Kaliveli. These shrimp farms, apart from adding pollution to the lake, may disturb the peacefully perching birds by increasing human activity in the surroundings.

(5) The resources of Kaliveli have been exploited traditionally for a long time. However, in the recent past the intensive fishing, both in Kaliveli and in Yedayanthitu estuary, seem to affect the piscifaunal diversity and thus indirectly it would affect the birds that feed on fishes.